Forestry: A Sustainable Approach

Forestry: A Sustainable Approach

Wesley Casey

STATES
ACADEMIC PRESS
www.statesacademicpress.com

Forestry: A Sustainable Approach
Wesley Casey
ISBN: 978-1-63989-211-2 (Hardback)

STATES
ACADEMIC PRESS

Published by States Academic Press,
109 South 5th Street,
Brooklyn, NY 11249, USA

Cataloging-in-Publication Data

Forestry : a sustainable approach / Wesley Casey.
p. cm.
Includes bibliographical references and index.
ISBN 978-1-63989-211-2
1. Forests and forestry. 2. Sustainable forestry. 3. Sustainable agriculture. I. Casey, Wesley.
SD131 .F67 2022
634.9--dc23

Table of Contents

Preface VII

Chapter 1 Introduction 1

a. Forest 1
b. Forestry 4
c. Sustainable Forestry 6

Chapter 2 Types of Forests 10

a. Evergreen Forest 10
b. Rainforest 17
c. Tropical and Subtropical Moist Broadleaf Forests 23
d. Tropical and Subtropical Dry Broadleaf Forests 25
e. Taiga 28
f. Freshwater Swamp Forest 42

Chapter 3 Sustainable Forest Management 45

a. Forest Management 45
b. Sustainable Forest Management 64
c. Forest Ecosystem Management 68
d. Ecoforestry 74
e. Forest Governance 77
f. Analog Forestry 78
g. Community Forestry 79
h. Continuous Cover Forestry 83

Chapter 4 Agroforestry 86

a. Forest Farming 89
b. Forest Gardening 93
c. Mycoforestry 96
d. Riparian Buffer 98

Chapter 5 Silviculture 103
a. Silvicultural System 131
b. Plant Nursery 137
c. Thinning 146
d. Ecological Thinning 149
e. Pruning 151
f. Clearcutting 156
g. Variable Retention 162
h. Coppicing 164

Chapter 6 Benefits of Sustainable Forestry 166
a. Carbon Sequestration 166
b. Biodiversity Conservation 170
c. Cleaner Air 185
d. Contribution to Agricultural Production 188
e. Protection of Water Quality 191

Chapter 7 Forest Conservation 196
a. Assisted Natural Regeneration 198
b. Reducing Emissions from Deforestation and Forest Degradation 200
c. Forest Restoration 219
d. Afforestation 224
e. Reforestation 230
f. Protective Measures for Forest Conservation 234

Permissions

Index

Preface

The purpose of this book is to help students understand the fundamental concepts of this discipline. It is designed to motivate students to learn and prosper. I am grateful for the support of my colleagues. I would also like to acknowledge the encouragement of my family.

The science and craft of creation, utilization, management, restoration and conservation of forests, woodlands and related resources for human and environmental benefits is called forestry. It is practiced in natural stands and plantations. It embraces a wide range of concerns which include erosion control, watershed management, recreation, landscape and community protection, biodiversity management, etc. This discipline develops and implements forest management plans. Forest genetic resources are an important factor which is considered under forestry. There are diverse branches of forestry such as agroforestry, ecoforestry, sustainable forestry, mycoforestry, permaforestry, energy forestry, community forestry, analog forestry, etc. This book is compiled in such a manner, that it will provide an in-depth knowledge about the theory and practice of forestry. Also included herein is a detailed explanation of its various concepts and applications. Coherent flow of topics, student-friendly language and extensive use of examples make this book an invaluable source of knowledge.

A foreword for all the chapters is provided below:

Chapter - Introduction

Forestry is the art and practice of managing, conserving and studying forests and plantations. Sustainable forestry deals with balancing the needs of the wildlife and the surrounding environment. The topics elaborated in this chapter will help in gaining a better perspective of sustainable forestry.

Chapter - Types of Forests

There are different types of forest depending on the forest land, species, biodiversity and surrounding climate. Some of them are rainforest, evergreen forest, freshwater swamp forest, tropical and sub-tropical forest, etc. This chapter sheds light on the various types of forests for an in-depth understanding of the subject.

Chapter - Sustainable Forest Management

Forest management deals with the economic, social, technical, environmental and ecological aspects of forest. Sustainable forest management maintains balance between all these aspects by following the principles of sustainable management. This chapter has been carefully written to provide an easy understanding of sustainable forest management.

Chapter – Agroforestry

Agroforestry is the combination of agriculture and forestry. It is a land-use system in which shrubs and trees are grown around pastureland to increase biodiversity and reduce erosion. Forest farming and gardening, mycoforestry, etc. fall within it. All the aspects related to agroforestry have been carefully analyzed in this chapter.

Chapter – Silviculture

The art, science and practice of controlling the structure, growth and quality of forests for meeting the needs of timber production is termed as silviculture. Ecological thinning, pruning, clearcutting, coppicing, etc. fall under its domain. This chapter closely examines silviculture to provide an extensive understanding of the subject.

Chapter – Benefits of Sustainable Forestry

Carbon sequestration, biodiversity conservation, conservation of soil, protection of water quality, improvement in air quality, reduction in soil erosion, etc. are some of the benefits of sustainable forestry. This chapter delves into these benefits of sustainable forestry for a thorough understanding of the subject.

Chapter – Forest Conservation

Forest conservation refers to the preservation and protection of forests for the benefit of future generations. Forest restoration, afforestation, reforestation, etc. are some methods used for the conservation of forests. This chapter discusses the concepts associated with forest conservation in detail.

Wesley Casey

1
Introduction

Forestry is the art and practice of managing, conserving and studying forests and plantations. Sustainable forestry deals with balancing the needs of the wildlife and the surrounding environment. The topics elaborated in this chapter will help in gaining a better perspective of sustainable forestry.

Forest

Forest is a complex ecological system in which trees are the dominant life-form.

Tree-dominated forests can occur wherever the temperatures rise above 10 °C (50 °F) in the warmest months and the annual precipitation is more than 200 mm (8 inches). They can develop under a variety of conditions within these climatic limits, and the kind of soil, plant, and animal life differs according to the extremes of environmental influences. In cool, high-latitude subpolar regions, forests are dominated by hardy conifers like pines, spruces, and larches. These taiga (boreal) forests have prolonged winters and between 250 and 500 mm (10 and 20 inches) of rainfall annually. In more temperate high-latitude climates, mixed forests of both conifers and broad-leaved deciduous trees predominate. Broad-leaved deciduous forests develop in middle-latitude climates, where there is an average temperature above 10 °C (50 °F) for at least six months every year and annual precipitation is above 400 mm (16 inches). A growing period of 100 to 200 days allows deciduous forests to be dominated by oaks, elms, birches, maples, beeches, and aspens. In the humid climates of the equatorial belt, tropical rainforests develop. There heavy rainfall supports evergreens that have broad leaves instead of needle leaves, as in cooler forests. In the lower latitudes of the Southern Hemisphere, the temperate deciduous forest reappears.

Forest types are distinguished from each other according to species composition (which develops in part according to the age of the forest), the density of tree cover, type of soils found there, and the geologic history of the forest region.

Boreal forest, Alaska, dominated by spruce trees (Picea).

Mixed evergreen and hardwood forest on the slopes of the Adirondack Mountains near Keene Valley.

Soil conditions are distinguished according to depth, fertility, and the presence of perennial roots. Soil depth is important because it determines the extent to which roots can penetrate into the earth and, therefore, the amount of water and nutrients available to the trees. The soil of taiga forests is sandy and quickly drained. Deciduous forests have brown soil, richer than sand in nutrients, and less porous. Rainforests and savanna woodlands have a soil layer rich in iron or aluminum, which give the soils either a reddish or yellowish cast. The amount of water available to the soil, and therefore available for tree growth, depends on the amount of annual rainfall. Water may be lost by evaporation from the surface or by leaf transpiration. Evaporation and transpiration also control the temperature of the air in forests, which is always slightly warmer in cold months and cooler in warm months than the air in surrounding regions.

The sun sets on a savanna in the African country of Kenya.

The density of tree cover influences the amount of both sunlight and rainfall reaching every forest layer. A full-canopied forest absorbs between 60 and 90 percent of available light, most of which is absorbed by the leaves for photosynthesis. The movement of rainfall into the forest is considerably influenced by leaf cover, which tends to slow the velocity of falling water, which penetrates down to the ground level by running down tree trunks or dripping from leaves. Water not absorbed by the tree roots for nutrition runs along root channels, so water erosion is therefore not a major factor in shaping forest topography.

Forests are among the most complex ecosystems in the world, and they exhibit extensive vertical stratification. Conifer forests have the simplest structure: A tree layer rising to about 98 feet (30 metres), a shrub layer that is spotty or even absent, and a ground layer covered with lichens, mosses, and liverworts. Deciduous forests are more complex; the tree canopy is divided into an upper and lower story, while rainforest canopies are divided into at least three strata. The forest floor in both of these forests consists of a layer of organic matter overlying mineral soil. The humus layer of tropical soils is affected by the high levels of heat and humidity, which quickly decompose whatever organic matter exists. Fungi on the soil surface play an important role in the availability and distribution of nutrients, particularly in the northern coniferous forests. Some species of fungi live in partnership with the tree roots, while others are parasitically destructive.

Animals that live in forests have highly developed hearing, and many are adapted for vertical movement through the environment. Because food other than ground plants is scarce, many ground-dwelling animals use forests only for shelter. In temperate forests, birds distribute plant seeds and insects aid in pollination, along with the wind. In tropical forests, fruit bats and birds effect pollination. The forest is nature's most efficient ecosystem, with a high rate of photosynthesis affecting both plant and animal systems in a series of complex organic relationships.

Sustainable Forest

A sustainable forest is a forest that is carefully managed so that as trees are felled they are replaced with seedlings that eventually grow into mature trees. This is a carefully and skilfully managed system. The forest is a working environment, producing wood products such as wood pulp for the paper/card industry and wood based materials for furniture manufacture and the construction industry. Great care is taken to ensure the safety of wildlife and to preserve the natural environment.

Sustainable forests are the result of a commonsense policy to replace trees that are felled so that forests continue to exist providing natural materials for us all.

A typical sustainable forest will contain trees of all ages and often different species of trees. As the trees mature they are felled and the natural wood is processed at saw mills.

Felled trees are replaced with seedlings. In this way the forest is constantly renewed.

Sustainable forests are not only productive in terms of the wood products they provide but also they are regarded as places that can be exploited by the leisure industry. Sustainable forests add greatly to the look of the natural environment and consequently attract tourists such as walkers and hikers.

Sustainable forests can only survive if they make a profit and provide employment, often for local people. The type of employment available will include work related directly to managing and maintaining the forests but also jobs related to the tourist industry. This will include the hotel and catering industry.

Forestry

Forestry is defined as the theory and practice of all that constitutes the creation, conservation and scientific management of forests and the utilization of their resources. It includes all thinking and all actions pertaining to creation and management of forests, including harvesting, marketing and utilization of all forest products and services. It includes not only management of existing forests but also the creation of new forests.

On the basis of Functions, all forest lands may be classified into:

- Protection Forests.
- National Forests.
- Village Forests.
- Tree Lands.

National Commission of Agriculture proposed that forests can be classified into:

Protection Forests

The practices of managing the forests for their protection function are called Protection forestry. In protection forestry, the object is to protect the site due to instability of terrain, nature of soil, geological formations, etc. Such areas where manipulation of the forest cover is not desirable may be classed as protection forests. The forests located on higher hill slopes, national parks and sanctuaries, preservation plots, biosphere or nature reserves and wilderness areas may be included under protection forests. The practice of forestry with a view to conserving flora, fauna, soil and water, increasing water yields, reducing floods and droughts, amelioration of climatic conditions, etc. is called Protection Forestry.

Production Forests

The practice of forestry with object of producing maximum quantity of timber, fuel wood and other forest produce is called Production Forestry. The production forestry can be further classified into:

- Commercial Forestry: Commercial forestry aims to get maximum production of timber, fuel wood and other forest products as a business enterprise.
- Industrial Forestry: Industrial forestry aims at producing raw material required for industry. In Production forestry, there is a greater concern for the production and economic returns.
- Mixed Quality Forests.
- Valuable Forests.
- Inaccessible Forests.

Social Forests

Social Forestry is the practice of forestry which aims at meeting the requirement of rural and urban population. The object of social forestry is to meet the basic needs of community aiming at bettering the conditions of living through:

- Meeting the fuel wood, fodder and small timber requirements.
- Protection of agricultural fields against wind.
- Meeting recreational needs.
- Maximizing production and increasing farm returns.

Various forms of Social forestry are:

- Community Forestry: The practice of forestry on lands outside the conventional forest area for the benefit of local population has been called Community forestry. Community forestry seeks the involvement of community in the creation and management of such forests.
- Farm Forestry: Farm Forestry is defined as the practice of forestry in all its aspect on farms of village lands generally integrated with other farm operations.
- Extension Forestry: Extension forestry which includes the activity of raising trees on farm lands, villages wastelands and community forest areas and on lands along the sides of roads, canal banks and railway lines. More recently, there has been emphasis on dynamic land use planning and efforts are made to maximise production on farmlands under agroforestry.
- Agro-Forestry: Agro-forestry has been defined as a sustainable land management

system which increases the yield of the land, combines the production of crops and forest plants and/or animals simultaneously or sequentially on the same unit of land and applies management practices that are compatible with the cultural practices of the local population.

- Recreational Forestry: More recently, there has been considerable demand for Recreational forestry, which is defined as the practice of forestry with the object of developing and maintaining forests of high scenic value. Recreational forests are being developed near towns and cities. The areas are being planted with flowering trees, shrubs and creepers to provide forest atmosphere near towns and cities.

Sustainable Forestry

"Sustainable" means to maintain, continue, and keep, while "forestry" is the science and art of managing forests. Thus, sustainable forestry is about caring for and managing forests to provide the natural resources, such as wood and clean water, we need now and in the future. It also means sustaining other things we value from the forest like wildlife habitat and beautiful landscapes. Sustainable forestry is concerned with all parts of the forest-trees, smaller plants, soils, wildlife, and water. It involves protecting forests from wildfire, pests, and diseases, and preserving forests that are unique or special.

Protecting the quality of water in the forest is critical to all aquatic and terrestrial life.

Sustainable forestry also puts an emphasis on people. People need forests for the resources they provide, and as a place to live or to make a living. Sustainable forestry ensures that future generations will have forests to meet their needs and values. As you can see, sustainable forestry is complex and can involve many things. Let's look at a few of these things more closely. You will also see how sustainable forestry can mean different things to different people.

Forest Management

To care for and manage a forest in a sustainable way it is necessary to use responsible management practices. These are often specifically adapted to each site. One of the most important practices is to look at whether the forest has enough natural seeds, seedlings, and tree sprouts (all called regeneration) to make a future forest. Excessive populations of white-tailed deer in a forest greatly reduce regeneration by eating young trees. Too many ferns or too little sunlight can also play a role. Many sustainable forestry practices can protect or encourage forest regeneration. They include putting up a fence to exclude deer, controlling weeds and other plants, and removing some trees to allow more sunlight to reach down into the forest.

When it comes to removing trees, or timber, from the forest, many practices can assure the forest's future. It is a good practice to think as much about which trees to leave as about which trees to cut. The trees left in the forest will continue to grow, occupying and regenerating the forest for many years to come. If they are chosen properly, the remaining trees can provide many of the same values and resources, and, perhaps, new ones in the future.

Other sustainable forestry practices include protecting forest streams and wet areas. Harvesting trees can disturb and expose soil in small areas. This is especially true on roads built for driving the machinery used to remove trees. The flow of rain or other water across, and under, roads must be carefully managed with culverts (large pipes) and proper road design. This helps prevent soil from washing into a stream or wet area. Too much soil in water, or sediment, is harmful to aquatic life. Trees and other vegetation left undisturbed adjacent to streams or wet areas can also prevent soil from entering streams. These areas are called buffer strips.

Browsing Concerns

Assuring that a forest has enough new trees to regenerate after a harvest is an important practice. Deer exclusion fencing (top) and tree tubes (bottom) can protect seedlings and stump sprouts from deer browsing.

Forest Health

Sustainable forestry is also about maintaining forest health. Many living and non-living things make a forest unhealthy. These include insect pests, diseases, wildfires, and extreme weather. Normally, the loss of individual or small groups of trees is not a big concern. When larger numbers of trees die, it may signal a forest health problem. Sustainable forestry seeks to promote forest health by reducing things which seriously damage the forest.

Gypsy Moths are just one of the pests that can impact forest health. They must be monitored and sometimes controlled to protect forests.

Caring for Special Sites

All forests are special, but some areas are more unique than others. It is a principle of sustainable forestry to protect unique areas that are not found elsewhere or are uncommon. One example is very old forests. Almost all of the 17 million acres of forests were clear-cut at least once by the early 1900s. However, a few small areas were never cut. Most of these old forests are currently preserved.

Pennsylvania has very few old growth forests left. Most of these are protected for future generations to enjoy and study.

Different Approaches

People view sustainable forestry in many different ways. To a forest landowner, sustainable forestry might mean selling timber and yet passing the land to an heir in good condition. To a logger (a person who cuts trees for a living) sustainable forestry might mean protecting the trees left in a forest, and constructing roads properly. To a professional forester, sustainable forestry might mean selecting appropriate forestry practices for a given site. All of these people want to sustain forests, even though each views the forest in different ways. Sustainable forestry is a broad but important concept. It requires concern and commitment on everyone's part.

High tech tools such as Global Positioning Systems (GPS), Geographic Information Systems (GIS), and computer modeling assist forestry professionals in the practice of sustainable forestry.

References

- Forest, science: britannica.com, Retrieved 02 January, 2020
- forestry-definition-and-types-of-forestry-1584: agriinfo.in, Retrieved 28 June, 2020
- Sustainable-forestry: extension.psu.edu, Retrieved 27 May, 2020
- Susenv-1: technologystudent.com, Retrieved 05 July, 2020

2

Types of Forests

There are different types of forest depending on the forest land, species, biodiversity and surrounding climate. Some of them are rainforest, evergreen forest, freshwater swamp forest, tropical and sub-tropical forest, etc. This chapter sheds light on the various types of forests for an in-depth understanding of the subject.

Evergreen Forest

Evergreen forest is a forest in which there is no complete, seasonal loss of leaves (i.e. trees shed old leaves and produce new ones partially, and sometimes throughout the year, rather than during particular periods). The trees can be softwood conifers or broad-leaved hardwoods. The distribution of these forests ranges through boreal, middle, and tropical latitudes. The northern coniferous forests and the equatorial rain forests are the most extensive evergreen forests. Evergreen coniferous forests predominate where the growing season is less than half the year; broad-leaved evergreen forest is found in regions that lack a prolonged dry season.

The tropical evergreen forests usually occur in areas receiving more than 200 cm of rainfall and having a temperature of 15 to 30 degrees Celsius. They occupy about seven per cent of the earth's land surface and habours more than half of the world's plants and animals. They are found mostly near the equator.

These forests are dense and multi-layered. They harbour many types of plants and animals. The trees are evergreen as there is no period of drought. They are mostly tall and hardwood type. Leaves are broad and give out excess water through evapo-transpiration.

Tidal or Mangrove Forests

Tidal forests are found along the edges of deltas of rivers in tropical and sub-tropical regions of the world. Such forests are salt tolerant and are flooded by seawater during high tide.

Some of these forests are dense and impenetrable. Only a limited number of plants are found in these evergreen forests.

Tropical evergreen rainforest biome provides optimum environmental conditions for the growth and development of plants and animals because it is characterized by high rainfall and temperature throughout the year. This biome is also called optimum biome because of uninterrupted supply of abundant moisture and water and heat throughout the year which ensures continuous and normal growth of plants. Normally, the evergreen rainforest biome extends between 10 °N and 10 °S latitudes.

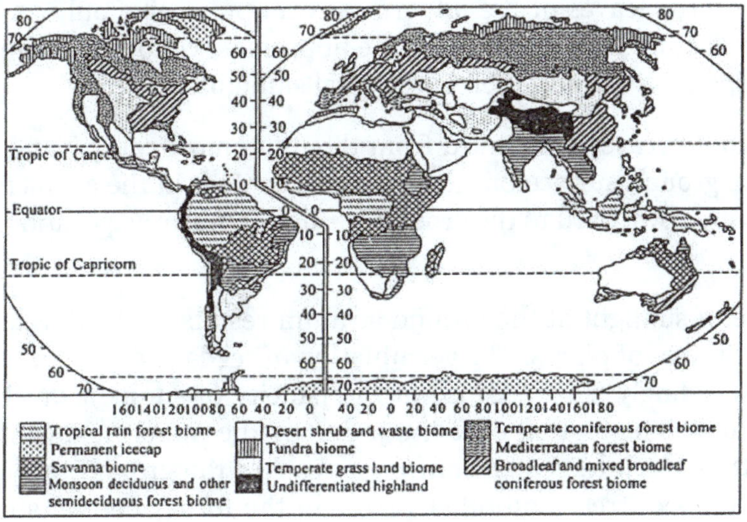

Major biomes of the world.

The maximum development of this biome has taken place in Amazon Basin (South America), Congo Basin (Africa) and Indo-Malaysian Region (mainly in Java, Sumatra, Borneo, Malaysia and Guinea). This biome has developed, infact, in true equatorial climatic region though in some areas the spatial coverage of this biome far extends beyond the equator. For example, it extends upto Vera Cruz in Mexico (about 19 °N) and upto 30 °S in South America.

Climate of Tropical Evergreen Rainforest Biome

The average annual rainfall of most of the areas of the tropical evergreen rainforest biome is about 2000 mm or even more. Except 2 or 3 months generally every month receives rainfall of at least 200 mm. The aridity during 8 few months certainly increases with increasing distance from the equator. There is high temperature throughout the year. Mean annual temperature is around 20 °C but the highest temperature of the year touches 30 °C. The sun is more or less overhead throughout the year.

There is very little difference between the lengths of day and night throughout the year. Annual range of temperature is around but the daily range of temperature varies between 5 °C and 10 °C. Most of the rainfall is received through convectional mechanism which yields heavy downpour through cumulonimbus clouds daily normally between 2 p.m. and 4 p.m.

The upper canopy of the tallest trees receives maximum sunlight whereas its amount decreases downward and it becomes minimum at the ground surface as there is complete darkness. This trend of progressively decreasing sunlight from the uppermost stratum to the ground stratum sets in keen competition among the various members of the vegetation community of the tropical evergreen rainforest biome for getting sunlight.

It may be pointed out that the uppermost canopy is also covered by climbers and creepers and therefore the leaves of the uppermost canopy receive only 25 percent of the total sunlight reaching the top surface of the uppermost canopy. The underlying areas of the tree canopy receive only 3 percent of total sunlight.

Wind speed also decreases downward from the uppermost canopy and it becomes almost zero at the ground surface but the moisture content in the air increases from the uppermost canopy downward to the ground surface because evaporation also decreases downward.

Almost absence of sunlight at the ground stratum results in the development of very few leaves and flowers of plants. The germination of seeds and growth of seedlings and saplings are exceedingly slow because of the paucity of sunlight on the forest floor. There is maximum interception of falling rains by the uppermost canopy of the forest cover and thus the intercepted rainwater reaches the ground through the leaves, branches and stems of trees and other plants in the form of aerial streamlets which allow maximum infiltration of rainwater in the ground surface.

The interception of rainwater by the uppermost forest canopy also allows much evaporation of intercepted rainwater because the rainwater is held at the outer surfaces of the leaves. Only one third of the total rainfall received at the uppermost forest canopy reaches the ground surface.

It is obvious that the vegetation has been controlled and determined by the typical characteristics of equatorial climate and in turn the vegetation has given birth to typical micro-climates in several strata of the vertical structure of the vegetation community of the tropical evergreen rainforest biome.

Species Composition of Tropical Evergreen Rainforest Biome

The tropical evergreen rainforest biome accounts for the largest number of plant species. Though there is almost uniformity and similarity in the life-forms and the structure of the plants in all parts of this biome but there is much variation in the composition of plant species. There are numerous species of plants in the various parts of the tropical evergreen rainforest biome.

For example, 6000 to 7000 species of flowering plants in the western Africa (Congo Basin); 20,000 species of flowering plants but quite different from the western Africa in Malaysia; 40,000 species of flowering plants in Brazil and 2000 species of flowering plants in Panama Canal Zone have been identified. It well be virtually impossible to present all the species of plants by names if all the species of all plants of the tropical evergreen rainforest biome become known to the botanists.

Tree is the most significant member of the tropical evergreen forests. The number of tree species is so large and their diversity is so great that one hectare of land in the equatorial region accounts for 40 to 100 species of trees. It may be pointed out that tree species account for 70 percent of the total plant species of the tropical evergreen rainforests. Creepers or climbers are the second important floral members of the rainforests.

These belong to the category of vines and range in form from fine string-like stems to massive cable-like forms. These creepers are so circuitous and highly irregular in form that it becomes difficult to find out their actual lengths. 'They ramble through the forest, scaling the highest emergent trees and frequently looping down to the ground and then ascending further sections of the forests. In the constant struggle for light, climbers have developed this mechanism for reaching the high insolation areas with the investment required for a large supporting biomass'.

The climbers, thus, so greatly bind several trees and plants together that the accessibility in the forest cover becomes almost zero. This typical structural form of the tropical evergreen rainforest has also affected the form of animal life. It may be pointed out that the tropical rainforests account for 90 percent of all climbing species.

Temperate Coniferous Forest

Temperate evergreen forests are found predominantly in areas with warm summers and cool winters, and vary enormously in their kinds of plant life. In some, needleleaf trees dominate, while others are home primarily to broadleaf evergreen trees or a mix of both tree types.

Temperate evergreen forests are common in the coastal areas of regions that have mild winters and heavy rainfall, or inland in drier climates or montane areas. Many species of trees inhabit these forests including pine, cedar, fir, and redwood.

The understory also contains a wide variety of herbaceous and shrub species. Temperate conifer forests sustain the highest levels of biomass in any terrestrial ecosystem and are notable for trees of massive proportions in temperate rainforest regions.

Structurally, these forests are rather simple, consisting of 2 layers generally: An overstory and understory. However, some forests may support a layer of shrubs. Pine forests support an herbaceous groundlayer that may be dominated by grasses and forbs that lend themselves to ecologically important wildfires. In contrast, the moist conditions found in temperate rain forests favor the dominance by ferns and some forbs.

Temperate rain forests only occur in 7 regions around the world - the Pacific Northwest, the Validivian forests of southwestern South America, the rain forests of New Zealand and Tasmania, the Northeastern Atlantic (small, isolated pockets in Ireland, Scotland, and Iceland), southwestern Japan, and those of the eastern Black Sea.

Forest communities dominated by huge trees (e.g., giant sequoia, Sequoiadendron gigantea; redwood, Sequoia sempervirens; mountain ash, Eucalyptus regnans), an unusual ecological phenomena, occur in western North America, southwestern South America, as well as in the Australasian region in such areas as southeastern Australia and northern New Zealand.

The Klamath-Siskiyou ecoregion of western North America harbors diverse and unusual assemblages and displays notable endemism for a number of plant and animal taxa.

Biodiversity Patterns

Most tree species and larger vertebrates have relatively widespread distributions; considerable local endemism and beta diversity occurs in some ecoregions in invertebrates, understory plants, and lichens, particularly in rain forests or on unusual soils; may have extremely diverse invertebrate faunas or herbaceous floras; altitudinal specialization occurs but is less pronounced than in the tropics.

Minimum Requirements

Disturbance regimes such as fire, windthrow, and epizootics can vary considerably within this major habitat type, but the extremes are typically of sufficient size and frequency as to make small patches of natural forest have only limited conservation value; many species highly specialized on late-successional forests; larger carnivores very wide-ranging with large home ranges; some species track resources that vary widely in space in time (e.g., epizootic outbreaks, fire events, cone production) requiring large natural landscapes.

Sensitivity to Disturbance

Larger carnivores highly sensitive to human activities including low intensity hunting; large number of species highly sensitive to logging and fragmentation of natural forests, particularly late-successional species; late-successional species and features typically regenerate slowly; many temperate forests require periodic fires to maintain successional processes and many species; exotic species can have extensive and significant impacts on natural forest communities.

Temperate Deciduous Forest

A temperate deciduous forest is a biome that has many deciduous trees which drop their leaves in the fall. These forests are also known as broad-leaf forests because the trees have wide, flat leaves. Temperate deciduous forests lie in the mid-latitude areas of the Earth, between the Arctic poles and the tropics. These biomes are exposed to warm and cold air masses, causing them to have four seasons: winter, spring, summer and fall. As winter approaches and daylight decreases, the production of chlorophyll in the leaves slows and eventually stops, revealing the bright red, yellow and orange colors we associate with fall. Temperate forests began to form in the Cenozoic Era about 65.5 million years ago when the Earth began to cool. Tropical and subtropical forests are the other types of deciduous forests.

Temperate Deciduous Forest Animals

Insects inhabiting temperate deciduous forests include ants, flies, bees, wasps, cicadas, walking sticks, moths, butterflies, dragonflies, mosquitoes and praying mantises. Frogs, toads, snakes and salamanders are some of the reptilian residents of the temperate deciduous forest. Common birds found in this biome include woodpeckers, robins, jays, cardinals, owls, turkeys, hawks and eagles. Smaller mammals in the temperate deciduous forests include rabbits, otters, monkeys, beavers, raccoons, porcupines and squirrels. Bears, white-tailed deer, moose, tigers, elephants, giraffes, leopards, pandas and humans are some of the larger mammals that live in this biome.

In the temperate deciduous forest, there is a food web that consists of several trophic (food) levels. Each trophic level has organisms that have the same role in the food web. They also share the same energy sources. At the lowest level, the soil is full of earthworms, bacteria and fungi (called decomposers), which generate nutrients for organisms in the level above. The primary producers in the second level use these nutrients and sunlight to produce energy through photosynthesis. The next level up has the primary consumers, mostly herbivores, which consume the primary producers in the trophic level below them. In turn, the primary consumers supply energy for the secondary consumers (carnivores and omnivores) in the next level. The uppermost level of the trophic food web is home to the top carnivores, also called tertiary consumers. In addition, there can be a quaternary trophic level in more complex food webs. The food web ends when the animals at the top have no natural predators.

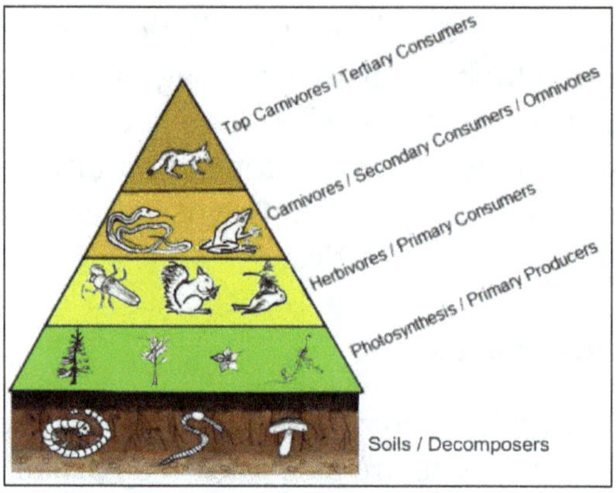

Temperate Deciduous Forest Plants

Temperate deciduous forests are home to trees such as oak, birch, beech, aspen, elm and maple. These tall trees form the forest canopy. Taller trees that stick up above the canopy are part of the emergent layer. After the canopy, the next layer down is the understory, which has smaller species of trees and younger trees that haven't yet reached their full height. The shrub layer is next and contains woody vegetation such as bushes and brambles that grow in places where enough light comes down through the canopy. After this, comes the herb layer, consisting of soft-stemmed (herbaceous) plants including wildflowers, ferns and grasses. These plants must be shade-tolerant to live this far down in the forest. The last layer is the forest floor. This area of the biome has rich soil from decayed leaves, twigs, moss and animal waste, also called litter; this is the recycling area of the temperate deciduous forest. Earthworms, bacteria, fungi and insects are resident here, and their physical and metabolic processes keep the recycling going. Recent research has shown that decreasing leaf litter results in rapid loss of carbon from the soil.

Temperate Deciduous Forest Climate

Temperate deciduous forests have temperatures ranging from -22°F to 86°F. The average annual rainfall is 30 to 60 inches, with added precipitation falling in the form of snow. Temperate deciduous forests need at least 120 days without frost. Temperate deciduous forest canopies allow some sunlight to penetrate to the forest floor. This creates more plant and animal diversity than is found in tropic or subtropical deciduous forests, both of which have dense canopies.

Rainforest

Ainforests are forests characterized by high and continuous rainfall, with annual rainfall in the case of tropical rainforests between 2.5 and 4.5 metres (98 and 177 in), and definitions varying by region for temperate rainforests. The monsoon trough, alternatively known as the intertropical convergence zone, plays a significant role in creating the climatic conditions necessary for the Earth's tropical rainforests: which are distinct from monsoonal areas of seasonal tropical forest.

Estimates vary from 40% to 75% of all biotic species are indigenous to the rainforests. There may be many millions of species of plants, insects and microorganisms still undiscovered in tropical rainforests. Tropical rainforests have been called the "jewels of the Earth" and the "world's largest pharmacy", because over one quarter of natural medicines have been discovered there. Rainforests are also responsible for 28% of the world's oxygen turnover, sometimes misnamed oxygen production, processing it through photosynthesis from carbon dioxide and consuming it through respiration.

The undergrowth in some areas of a rainforest can be restricted by poor penetration of sunlight to ground level. If the leaf canopy is destroyed or thinned, the ground beneath is soon colonized by a dense, tangled growth of vines, shrubs and small trees, called a jungle. The term *jungle* is also sometimes applied to tropical rainforests generally.

Rainforests as well as endemic rainforest species are rapidly disappearing due to deforestation, the resulting habitat loss and pollution of the atmosphere.

Tropical

Tropical rainforests are characterized by a warm and wet climate with no substantial dry season: typically found within 10 degrees north and south of the equator. Mean monthly temperatures exceed 18 °C (64 °F) during all months of the year. Average annual rainfall is no less than 168 cm (66 in) and can exceed 1,000 cm (390 in) although it typically lies between 175 cm (69 in) and 200 cm (79 in).

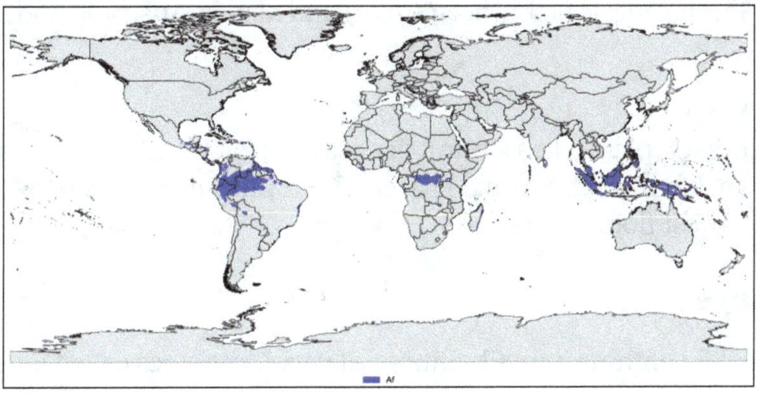

Worldwide tropical rainforest climate zones.

Many of the world's tropical forests are associated with the location of the monsoon trough, also known as the intertropical convergence zone. The broader category of tropical moist forests are located in the equatorial zone between the Tropic of Cancer and Tropic of Capricorn. Tropical rainforests exist in Southeast Asia (from Myanmar (Burma)) to the Philippines, Malaysia, Indonesia, Papua New Guinea and Sri Lanka; also in Sub-Saharan Africa from the Cameroon to the Congo (Congo Rainforest), South America (e.g. the Amazon rainforest), Central America (e.g. Bosawás, the southern Yucatán Peninsula-El Peten-Belize-Calakmul), Australia, and on Pacific Islands (such as Hawai'i). Tropical forests have been called the "Earth's lungs", although it is now known that rainforests contribute little net oxygen addition to the atmosphere through photosynthesis.

Temperate

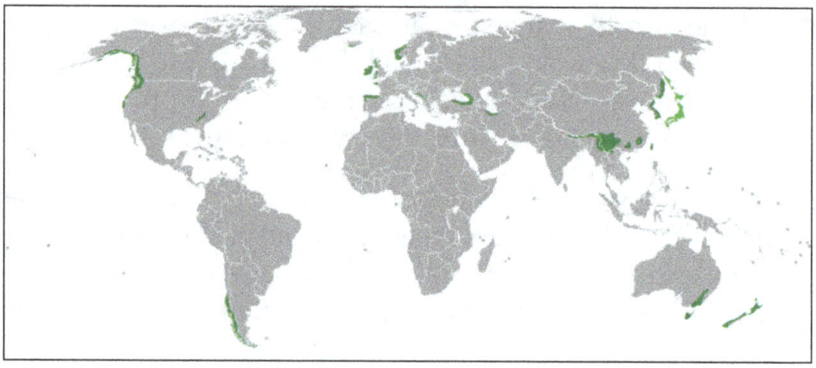

General distribution of temperate rainforests.

Tropical forests cover a large part of the globe, but temperate rainforests only occur in few regions around the world. Temperate rainforests are rainforests in temperate regions. They occur in North America (in the Pacific Northwest in Alaska, British Columbia, Washington, Oregon and California), in Europe (parts of the British Isles such as the coastal areas of Ireland and Scotland, southern Norway, parts of the western Balkans along the Adriatic coast, as well as in Galicia and coastal areas of the eastern Black Sea, including Georgia and coastal Turkey), in East Asia (in southern China, Highlands of Taiwan, much of Japan and Korea, and on Sakhalin Island and the adjacent Russian Far East coast), in South America (southern Chile) and also in Australia and New Zealand.

Layers

A tropical rainforest typically has a number of layers, each with different plants and animals adapted for life in that particular area. Examples include the emergent, canopy, understory and forest floor layers.

Emergent Layer

The emergent layer contains a small number of very large trees called emergents, which grow above the general canopy, reaching heights of 45–55 m, although on occasion a few species will grow to 70–80 m tall. They need to be able to withstand the hot temperatures and strong winds that occur above the canopy in some areas. Eagles, butterflies, bats and certain monkeys inhabit this layer.

The canopy at the Forest Research Institute Malaysia showing crown shyness.

Canopy Layer

The canopy layer contains the majority of the largest trees, typically 30 metres (98 ft) to 45 metres (148 ft) tall. The densest areas of biodiversity are found in the forest canopy, a more or less continuous cover of foliage formed by adjacent treetops. The canopy, by some estimates, is home to 50 percent of all plant species. Epiphytic plants attach to trunks and branches, and obtain water and minerals from rain and debris that collects on the supporting plants. The fauna is similar to that found in the emergent layer, but more diverse. A quarter of all insect species are believed to exist in the rainforest canopy. Scientists have long suspected the richness of the canopy as a habitat, but have only recently developed practical methods of exploring it. As long ago as 1917, naturalist William Beebe declared that "another continent of life remains to be discovered, not upon the Earth, but one to two hundred feet above it, extending over thousands of square miles." True exploration of this habitat only began in the 1980s, when scientists developed methods to reach the canopy, such as firing ropes into the trees using crossbows. Exploration of the canopy is still in its infancy, but other methods include the use of balloons and airships to float above the highest branches and the building of cranes and walkways planted on the forest floor. The science of accessing tropical forest canopy using airships or similar aerial platforms is called dendronautics.

Understory Layer

The understory or understorey layer lies between the canopy and the forest floor. It is home to a number of birds, snakes and lizards, as well as predators such as jaguars, boa constrictors and leopards. The leaves are much larger at this level and insect life is abundant. Many seedlings that will grow to the canopy level are present in the understory. Only about 5% of the sunlight shining on the rainforest canopy reaches the understory. This layer can be called a *shrub layer*, although the shrub layer may also be considered a separate layer.

Forest Floor

Rainforest in the Blue Mountains, Australia.

The forest floor, the bottom-most layer, receives only 2% of the sunlight. Only plants adapted to low light can grow in this region. Away from riverbanks, swamps and clearings, where dense undergrowth is found, the forest floor is relatively clear of vegetation because of the low sunlight penetration. It also contains decaying plant and animal matter, which disappears quickly, because the warm, humid conditions promote rapid decay. Many forms of fungi growing here help decay the animal and plant waste.

Flora and Fauna

More than half of the world's species of plants and animals are found in the rainforest. Rainforests support a very broad array of fauna, including mammals, reptiles, birds and invertebrates. Mammals may include primates, felids and other families. Reptiles include snakes, turtles, chameleons and other families; while birds include such families as vangidae and Cuculidae. Dozens of families of invertebrates are found in rainforests. Fungi are also very common in rainforest areas as they can feed on the decomposing remains of plants and animals.

A Kermode bear from the Great Bear Rainforest.

The great diversity in rainforest species is in large part the result of diverse and numerous physical refuges, i.e. places in which plants are inaccessible to many herbivores, or in which animals can hide from predators. Having numerous refuges available also results in much higher total biomass than would otherwise be possible.

Soils

Despite the growth of vegetation in a tropical rainforest, soil quality is often quite poor. Rapid bacterial decay prevents the accumulation of humus. The concentration of iron and aluminium oxides by the laterization process gives the oxisols a bright red colour and sometimes produces mineral deposits such as bauxite. Most trees have roots near the surface, because there are insufficient nutrients below the surface; most of the trees' minerals come from the top layer of decomposing leaves and animals. On younger substrates, especially of volcanic origin, tropical soils may be quite fertile. If rainforest trees are cleared, rain can accumulate on the exposed soil surfaces, creating run-off and beginning a process of soil erosion. Eventually streams and rivers form and flooding becomes possible. There are several reasons for the poor soil quality: First is that the soil is highly acidic. The roots of plants rely on an acidity difference between the roots and the soil in order to absorb nutrients. When the soil is acidic, there is little difference, and therefore little absorption of nutrients from the soil. Second, the type of clay particles present in tropical rainforest soil has a poor ability to trap nutrients and stop them from washing away. Even if humans artificially add nutrients to the soil, the nutrients mostly wash away and are not absorbed by the plants. Thirdly, the type of clay particles present in tropical rainforest soil has a poor ability to trap nutrients and stop them from washing away. Even if humans artificially added nutrients to the soil, the nutrients would still mostly wash away and not absorbed by the plants. Finally, these soils are poor due to the high volume of rain in tropical rainforests washes nutrients out of the soil more quickly than in other climates.

Effect on Global Climate

A natural rainforest emits and absorbs vast quantities of carbon dioxide. On a global scale, long-term fluxes are approximately in balance, so that an undisturbed rainforest would have a small net impact on atmospheric carbon dioxide levels, though they may have other climatic effects (on cloud formation, for example, by recycling water vapour). No rainforest today can be considered to be undisturbed. Human-induced deforestation plays a significant role in causing rainforests to release carbon dioxide, as do other factors, whether human-induced or natural, which result in tree death, such as burning and drought. Some climate models operating with interactive vegetation predict a large loss of Amazonian rainforest around 2050 due to drought, forest dieback and the subsequent release of more carbon dioxide. Five million years from now, the Amazon rainforest may long since have dried and transformed itself into savannah, killing itself in the process (changes such as this may happen even if all human deforestation activity ceases overnight).

Human Uses

Tropical rainforests provide timber as well as animal products such as meat and hides. Rainforests also have value as tourism destinations and for the ecosystem services provided. Many foods originally came from tropical forests, and are still mostly grown on plantations in regions that were formerly primary forest. Also, plant-derived medicines are commonly used for fever, fungal infections, burns, gastrointestinal problems, pain, respiratory problems, and wound treatment. At the same time, rainforests are usually not used sustainably by non-native peoples but are being exploited or removed for agricultural purposes.

Aerial view of the Amazon rainforest, taken from a plane.

Deforestation

Tropical and temperate rainforests have been subjected to heavy legal and illegal logging for their valuable hardwoods and agricultural clearance (slash-and-burn, clearcutting) throughout the 20th century and the area covered by rainforests around the world is shrinking. Biologists have estimated that large numbers of species are being driven to extinction due to the removal of habitat with destruction of the rainforests.

Satellite photograph of the haze above
Borneo and Sumatra on 24 September 2015.

Another factor causing the loss of rainforest is expanding urban areas. Littoral rainforest growing along coastal areas of eastern Australia is now rare due to ribbon development to accommodate the demand for seachange lifestyles.

Forests are being destroyed at a rapid pace. Almost 90% of West Africa's rainforest has been destroyed. Since the arrival of humans, Madagascar has lost two thirds of its original rainforest. At present rates, tropical rainforests in Indonesia would be logged out in 10 years and Papua New Guinea in 13 to 16 years. According to Rainforest Rescue, an important reason for the increasing deforestation rate, especially in Indonesia, is the expansion of oil palm plantations to meet growing demand for cheap vegetable fats and biofuels. In Indonesia, palm oil is already cultivated on nine million hectares and, together with Malaysia, the island nation produces about 85 percent of the world's palm oil.

Several countries, notably Brazil, have declared their deforestation a national emergency. Amazon deforestation jumped by 69% in 2008 compared to 2007's twelve months, according to official government data.

By one estimate, for every acre of rainforest cut down each year, more than 50 acres of new forest are growing in the tropics." The new forest includes secondary forest on former farmland and so-called degraded forest.

Tropical and Subtropical Moist Broadleaf Forests

Tropical and subtropical moist broadleaf forests (TSMF), also known as tropical moist forests, are a tropical and subtropical forest habitat type defined by the World Wide Fund for Nature. The habitat type is sometimes known as jungle.

Rainforest lining a river bank, Cameroon.

TSMF are generally found in large, discontinuous patches centered on the equatorial belt and between the Tropics of Cancer and Capricorn, TSMF are characterized by low variability in annual temperature and high levels of rainfall (>200 centimetres (79 in) annually). Forest composition is dominated by semi-evergreen and evergreen deciduous tree species. These trees number in the thousands and contribute to the highest levels of species diversity in any terrestrial major habitat type. In general, biodiversity is highest in the forest canopy. The canopy can be divided into five layers: overstory canopy with emergent crowns, a medium layer of canopy, lower canopy, shrub level, and finally understory.

These forests are home to more species than any other terrestrial ecosystem: Half of the world's species may live in these forests, where a square kilometer may be home to more than 1,000 tree species. These forests are found around the world, particularly in the Indo-Malayan Archipelago, the Amazon Basin, and the African Congo Basin.

A perpetually warm, wet climate promotes more explosive plant growth than in any other environment on Earth. A tree here may grow over 23 metres (75 ft) in height in just 5 years. From above, the forest appears as an unending sea of green, broken only by occasional, taller "emergent" trees. These towering emergents are the realm of hornbills, toucans, and the harpy eagle.

The canopy is home to many of the forest's animals, including apes and monkeys. Below the canopy, a lower understory hosts to snakes and big cats. The forest floor, relatively clear of undergrowth due to the thick canopy above, is prowled by other animals such as gorillas and deer.

All levels of these forests contain an unparalleled diversity of invertebrate species, including New Guinea's stick insects and butterflies that can grow over 30 centimetres (1 ft) in length.

Many forests are being cleared for farmland, while others are subject to large-scale commercial logging. An area the size of Ireland is destroyed every few years.

Types

The biome includes several types of forests:

- Lowland equatorial evergreen rain forests, commonly known as tropical rainforests, are forests which receive high rainfall (tropical rainforest climate with more than 2000 mm, or 80 inches, annually) throughout the year. These forests occur in a belt around the equator, with the largest areas in the Amazon basin of South America, the Congo basin of central Africa, and parts of the Malay Archipelago. About half of the world's tropical rainforests are in the South American countries of Brazil and Peru. Rainforests now cover less than 6% of Earth's land surface. Scientists estimate that more than half of all the world's plant and animal species live in tropical rainforests.
- Tropical seasonal forests, also known as moist deciduous, monsoon or semi-evergreen (mixed) seasonal forests, have a monsoon or wet savannah climates (as in the Köppen climate classification): receiving high overall rainfall with a warm summer wet season and (often) a cooler winter dry season. Some trees in these

forests drop some or all of their leaves during the winter dry season. These forests are found in South Florida, parts of South America, in Central America and around the Caribbean, in coastal West Africa, parts of the Indian subcontinent, and across much of Indochina.

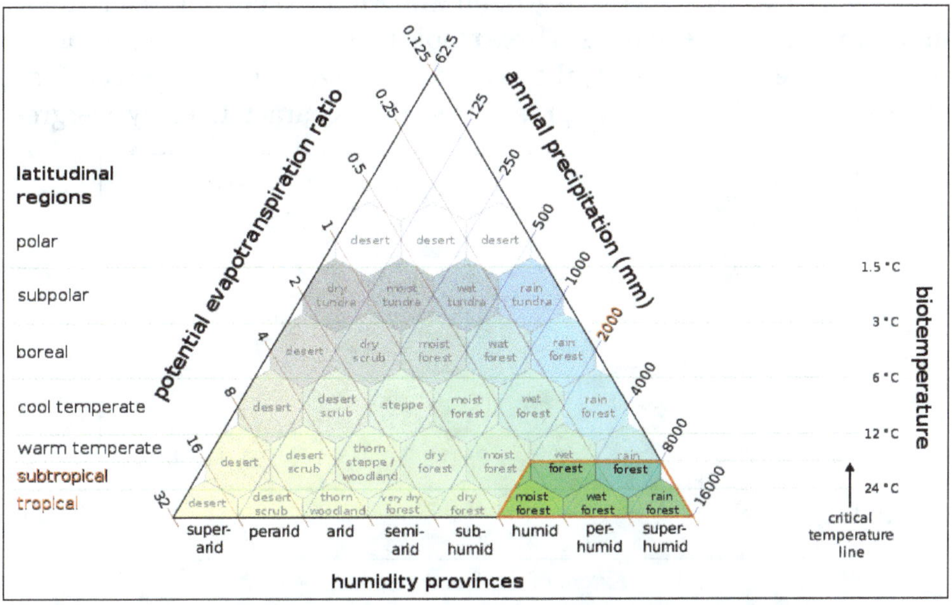

Tropical and subtropical moist forests (TSMF) as shown within the Holdridge Life Zones classification scheme, and includes moist forests, wet forests, and rainforests.

- Montane rain forests are found in cooler-climate mountainous areas. Those with elevations high enough to regularly encounter low-level cloud cover are known as cloud forests.
- Flooded forests, including freshwater swamp forests and peat swamp forests.

Tropical and Subtropical Dry Broadleaf Forests

The tropical and subtropical dry broadleaf forest is a habitat type defined by the World Wide Fund for Nature and is located at tropical and subtropical latitudes. Though these forests occur in climates that are warm year-round, and may receive several hundred centimeters of rain per year, they have long dry seasons which last several months and vary with geographic location. These seasonal droughts have great impact on all living things in the forest.

Deciduous trees predominate in most of these forests, and during the drought a leafless period occurs, which varies with species type. Because trees lose moisture through their leaves, the shedding of leaves allows trees such as teak and mountain ebony to conserve water during dry periods. The newly bare trees open up the canopy layer, enabling sunlight to reach ground level and facilitate the growth of thick underbrush. Trees on moister sites and those with access to ground water tend to be evergreen. Infertile sites also tend to support evergreen trees. Three tropical dry broadleaf forest ecoregions, the East Deccan dry evergreen forests, the Sri Lanka dry-zone dry evergreen forests, and the Southeastern Indochina dry evergreen forests, are characterized by evergreen trees.

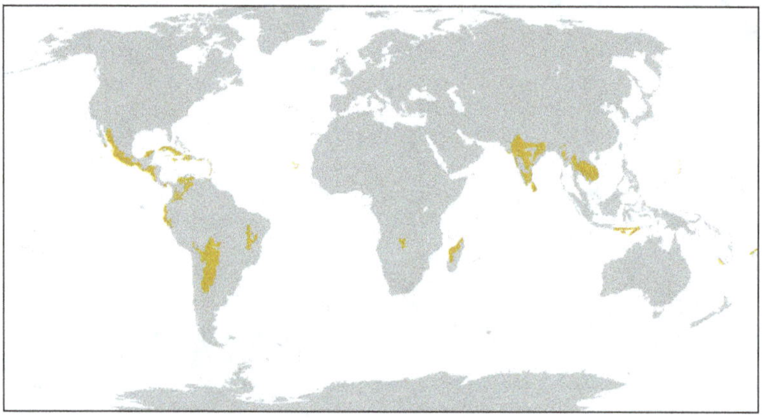

Extent of dry forest regions.

Though less biologically diverse than rainforests, tropical dry forests are home to a wide variety of wildlife including monkeys, deer, large cats, parrots, various rodents, and ground dwelling birds. Mammalian biomass tends to be higher in dry forests than in rain forests, especially in Asian and African dry forests. Many of these species display extraordinary adaptations to the difficult climate.

This biome is alternately known as the tropical bane forest biome or the tropical and subtropical deciduous forest biome.

Geographical Variation

Trinidad and Tobago dry forest on Chacachacare showing the dry-season deciduous nature of the vegetation

Dry forests tend to exist in the drier areas north and south of the tropical rainforest belt, south or north of the subtropical deserts, generally in two bands: one between 10° and 20 °N latitude and the other between 10° and 20 °S latitude. The most diverse dry forests in the world occur in southern Mexico and in the Bolivian lowlands. The dry forests of the Pacific Coast of northwestern South America support a wealth of unique species due to their dry climate. The Maputaland-Pondoland bushland and thickets along the east coast of South Africa are diverse and support many endemic species. The dry forests of central India and Indochina are notable for their diverse large vertebrate faunas. Madagascar dry deciduous forests and New Caledonia dry forests are also highly distinctive for a wide range of taxa and at higher taxonomic levels. Trees use underground water during the dry seasons.

Biodiversity Patterns and Requirements

Subtropical semi-evergreen seasonal forest in Doi Inthanon National Park, Northern Thailand, at the end of the dry season.

Species tend to have wider ranges than moist forest species, although in some regions many species do display highly restricted ranges; most dry forest species are restricted to tropical dry forests, particularly in plants; beta diversity and alpha diversity high but typically lower than adjacent moist forests.

Effective conservation of dry broadleaf forests requires the preservation of large and continuous areas of forest. Large natural areas are required to maintain larger predators and other vertebrates, and to buffer sensitive species from hunting pressure. The persistence of riparian forests and water sources is critical for many dry forest species. Large swathes of intact forest are required to allow species to recover from occasional large events, like forest fires.

Dry forests are highly sensitive to excessive burning and deforestation; overgrazing and exotic species can also quickly alter natural communities; restoration is possible but challenging, particularly if degradation has been intense and persistent.

Taiga

Taiga, generally referred to in North America as boreal forest or snow forest, is a biome characterized by coniferous forests consisting mostly of pines, spruces, and larches.

The taiga or boreal forest is the world's largest land biome. In North America, it covers most of inland Canada, Alaska, and parts of the northern contiguous United States. In Eurasia, it covers most of Sweden, Finland, much of Norway and Estonia, some of the Scottish Highlands, some lowland/coastal areas of Iceland, much of Russia from Karelia in the west to the Pacific Ocean (including much of Siberia), and areas of northern Kazakhstan, northern Mongolia, and northern Japan (on the island of Hokkaidō). However, the main tree species, the length of the growing season and summer temperatures vary. For example, the taiga of North America mostly consists of spruces; Scandinavian and Finnish taiga consists of a mix of spruce, pines and birch; Russian taiga has spruces, pines and larches depending on the region, while the Eastern Siberian taiga is a vast larch forest.

A different use of the term taiga is often encountered in the English language, with "boreal forest" used in the United States and Canada to refer to only the more southerly part of the biome, while "taiga" is used to describe the more barren areas of the northernmost part of the biome approaching the tree line and the tundra biome. Hoffman discusses the origin of this differential use in North America and why it is an inappropriate differentiation of the Russian term. Although at high elevations taiga grades into alpine tundra through Krummholz, it is not exclusively an alpine biome; and unlike subalpine forest, much of taiga is lowlands.

White spruce taiga in the Alaska Range.

Climate and Geography

Taiga is the world's largest land biome (depending on how one defines a biome, it could also be considered the second-largest, after deserts and xeric shrublands), covering 17 million square kilometres (6.6 million square miles) or 11.5% of the Earth's land area. The largest areas are located in Russia and Canada. The taiga is the terrestrial biome with the lowest annual average temperatures after the tundra and permanent ice caps. Extreme winter minimums in the northern taiga are typically lower than those of the tundra. The lowest reliably recorded temperatures in the Northern Hemisphere were recorded in the taiga of northeastern Russia. The taiga or boreal forest has a subarctic climate with very large temperature range between seasons, but the long and cold winter is the dominant feature. This climate is classified as *Dfc, Dwc, Dsc, Dfd* and *Dwd* in the Köppen climate classification scheme, meaning that the short summer (24 h average 10 °C (50 °F) or more) lasts 1–3 months and always less than 4 months. In Siberian taiga the average temperature of the coldest month is between −6 °C (21 °F) and −50 °C (−58 °F). There are also some much smaller areas grading towards the oceanic *Cfc* climate with milder winters, whilst the extreme south and (in Eurasia) west of the taiga reaches into humid continental climates (*Dfb, Dwb*) with longer summers. The mean annual temperature generally varies from −5 to 5 °C (23 to 41 °F), but there are taiga areas in eastern Siberia and interior Alaska-Yukon where the mean annual reaches down to −10 °C (14 °F). According to some sources, the boreal forest grades into a temperate mixed forest when mean annual temperature reaches about 3 °C (37 °F). Discontinuous permafrost is found in areas with mean annual temperature below freezing (0 °C; 32 °F), whilst in the *Dfd* and *Dwd* climate zones continuous permafrost occurs and restricts growth to very shallow-rooted trees like Siberian larch. The winters, with average temperatures below freezing, last five to seven months. Temperatures vary from −54 to 30 °C (−65 to 86 °F) throughout the whole year. The summers, while short, are generally warm and humid. In much of the taiga, −20 °C (−4 °F) would be a typical winter day temperature and 18 °C (64 °F) an average summer day.

The taiga in the river valley near Verkhoyansk, Russia, at 67 °N, experiences the coldest winter temperatures in the northern hemisphere, but the extreme continentality of the climate gives an average daily high of 22 °C (72 °F) in July.

The growing season, when the vegetation in the taiga comes alive, is usually slightly longer than the climatic definition of summer as the plants of the boreal biome have a lower threshold to trigger growth. In Canada, Scandinavia and Finland, the growing season is often estimated by using the period of the year when the 24-hour average temperature is +5 °C (41 °F) or more.

Boreal forest near Shovel Point in Tettegouche State Park, along the northern shore of Lake Superior in Minnesota.

For the Taiga Plains in Canada, growing season varies from 80 to 150 days, and in the Taiga Shield from 100 to 140 days. Some sources claim 130 days growing season as typical for the taiga. Other sources mention that 50–100 frost-free days are characteristic. Data for locations in southwest Yukon gives 80–120 frost-free days. The closed canopy boreal forest in Kenozersky National Park near Plesetsk, Arkhangelsk Province, Russia, on average has 108 frost-free days. The longest growing season is found in the smaller areas with oceanic influences; in coastal areas of Scandinavia and Finland, the growing season of the closed boreal forest can be 145–180 days.

The shortest growing season is found at the northern taiga–tundra ecotone, where the northern taiga forest no longer can grow and the tundra dominates the landscape when the growing season is down to 50–70 days, and the 24-hr average of the warmest month of the year usually is 10 °C (50 °F) or less. High latitudes mean that the sun does not rise far above the horizon, and less solar energy is received than further south. But the high latitude also ensures very long summer days, as the sun stays above the horizon nearly 20 hours each day, with only around 6 hours of daylight occurring in the dark winters, depending on latitude. The areas of the taiga inside the Arctic Circle have midnight sun in mid-summer and polar night in mid-winter.

Lakes and other water bodies are common in the taiga. The Helvetinjärvi National Park, Finland, is situated in the closed canopy taiga (mid-boreal to south-boreal) with mean annual temperature of 4 °C (39 °F).

The taiga experiences relatively low precipitation throughout the year (generally 200–750 mm (7.9–29.5 in) annually, 1,000 mm (39 in) in some areas), primarily as rain during the summer months, but also as fog and snow. This fog, especially predominant in low-lying areas during and after the thawing of frozen Arctic seas, means that sunshine is not abundant in the taiga even during the long summer days. As evaporation is consequently low for most of the year, precipitation exceeds evaporation, and is sufficient to sustain the dense vegetation growth including large trees. (In the steppe biome, often found south of taiga in the northern hemisphere, evapotranspiration exceeds precipitation, restricting vegetation to mostly grasses.)

Snow may remain on the ground for as long as nine months in the northernmost extensions of the taiga ecozone.

In general, taiga grows to the south of the 10 °C (50 °F) July isotherm, but occasionally as far north as the 9 °C (48 °F) July isotherm. Rich in spruces, Scots pines in the western Siberian plain, the taiga is dominated by larch in Eastern Siberia, before returning to its original floristic richness on the Pacific shores. Two deciduous trees mingle throughout southern Siberia: birch and Populus tremula.

Late September in the fjords near Narvik, Norway. This oceanic part of the forest can see more than 1,000 mm (39 in) precipitation annually and has warmer winters than the vast inland taiga.

The southern limit is more variable, depending on rainfall; taiga may be replaced by forest steppe south of the 15 °C (59 °F) July isotherm where rainfall is very low, but more typically extends south to the 18 °C (64 °F) July isotherm, and locally where rainfall is higher (notably in eastern Siberia and adjacent Outer Manchuria) south to the 20 °C (68 °F) July isotherm. In these warmer areas the taiga has higher species diversity, with more warmth-loving species such as Korean pine, Jezo spruce, and Manchurian fir, and merges gradually into mixed temperate forest or, more locally (on the Pacific Ocean coasts of North America and Asia), into coniferous temperate rainforests where oak and hornbeam appear and join the conifers, birch and Populus tremula.

Yukon, Canada. Several of the world's longest rivers go through the taiga, including Ob, Yenisei, Lena, and Mackenzie.

The area currently classified as taiga in Europe and North America (except Alaska) was recently glaciated. As the glaciers receded they left depressions in the topography that have since filled with water, creating lakes and bogs (especially muskeg soil) found throughout the taiga.

In Sweden the taiga is associated with the Norrland terrain.

Soils

Taiga soil tends to be young and poor in nutrients. It lacks the deep, organically enriched profile present in temperate deciduous forests. The thinness of the soil is due largely to the cold, which hinders the development of soil and the ease with which plants can use its nutrients. Fallen leaves and moss can remain on the forest floor for a long time in the cool, moist climate, which limits their organic contribution to the soil; acids from evergreen needles further leach the soil, creating spodosol, also known as podzol. Since the soil is acidic due to the falling pine needles, the forest floor has only lichens and some mosses growing on it. In clearings in the forest and in areas with more boreal deciduous trees, there are more herbs and berries growing. Diversity of soil organisms in the boreal forest is high, comparable to the tropical rainforest.

Flora

Since North America and Asia used to be connected by the Bering land bridge, a number of animal and plant species (more animals than plants) were able to colonize both continents and are distributed throughout the taiga biome. Others differ regionally, typically with each genus having several distinct species, each occupying different regions of the taiga. Taigas also have some small-leaved deciduous trees like birch, alder, willow, and poplar; mostly in areas escaping the most extreme winter cold. However, the Dahurian larch tolerates the coldest winters in the Northern Hemisphere in eastern Siberia. The very southernmost parts of the taiga may have trees such as oak, maple, elm and lime scattered among the conifers, and there is usually a gradual transition into a temperate mixed forest, such as the eastern forest-boreal transition of eastern Canada. In the interior of the continents with the driest climate, the boreal forests might grade into temperate grassland.

Boreal forest near Lake Baikal in Russia.

There are two major types of taiga. The southern part is the closed canopy forest, consisting of many closely spaced trees with mossy ground cover. In clearings in the forest, shrubs and wildflowers are common, such as the fireweed. The other type is the lichen woodland or sparse taiga, with trees that are farther-spaced and lichen ground cover; the latter is common in the northernmost taiga. In the northernmost taiga the forest cover is not only more sparse, but often stunted in growth form; moreover, ice pruned asymmetric black spruce (in North America) are often seen, with diminished foliage on the windward side. In Canada, Scandinavia and Finland, the boreal forest is usually divided into three subzones: The high boreal (north boreal) or taiga zone; the middle boreal (closed forest); and the southern boreal, a closed canopy boreal forest with some scattered temperate deciduous trees among the conifers, such as maple, elm and oak. This southern boreal forest experiences the longest and warmest growing season of the biome, and in some regions (including Scandinavia, Finland and western Russia) this subzone is commonly used for agricultural purposes. The boreal forest is home to many types of berries; some are confined to the southern and middle closed boreal forest (such as wild strawberry and partridgeberry); others grow in most areas of the taiga (such as cranberry and cloudberry), and some can grow in both the taiga and the low arctic (southern part of) tundra (such as bilberry, bunchberry and lingonberry).

Taiga spruce forest in the Kenai National Wildlife Refuge, Alaska. Trees in this environment tend to grow closer to the trunk and not "bush out" in the normal manner of spruce trees.

The forests of the taiga are largely coniferous, dominated by larch, spruce, fir and pine. The woodland mix varies according to geography and climate so for example the Eastern Canadian forests ecoregion of the higher elevations of the Laurentian Mountains and the northern Appalachian Mountains in Canada is dominated by balsam fir *Abies balsamea*, while further north the Eastern Canadian Shield taiga of northern Quebec and Labrador is notably black spruce *Picea mariana* and tamarack larch *Larix laricina*.

Evergreen species in the taiga (spruce, fir, and pine) have a number of adaptations specifically for survival in harsh taiga winters, although larch, which is extremely cold-tolerant, is deciduous. Taiga trees tend to have shallow roots to take advantage of the thin soils, while many of them seasonally alter their biochemistry to make them more resistant to freezing, called "hardening". The narrow conical shape of northern conifers, and their downward-drooping limbs, also help them shed snow.

Moss (*Ptilium crista-castrensis*) cover on the floor of taiga.

Because the sun is low in the horizon for most of the year, it is difficult for plants to generate energy from photosynthesis. Pine, spruce and fir do not lose their leaves seasonally and are able to photosynthesize with their older leaves in late winter and spring when light is good but temperatures are still too low for new growth to commence. The adaptation of evergreen needles limits the water lost due to transpiration and their dark green color increases their absorption of sunlight. Although precipitation is not a limiting factor, the ground freezes during the winter months and plant roots are unable to absorb water, so desiccation can be a severe problem in late winter for evergreens.

Although the taiga is dominated by coniferous forests, some broadleaf trees also occur, notably birch, aspen, willow, and rowan. Many smaller herbaceous plants, such as ferns and occasionally ramps grow closer to the ground. Periodic stand-replacing wildfires (with return times of between 20–200 years) clear out the tree canopies, allowing sunlight to invigorate new growth on the forest floor. For some species, wildfires are a necessary part of the life cycle in the taiga; some, e.g. jack pine have cones which only open to release their seed after a fire, dispersing their seeds onto the newly cleared ground; certain species of fungi (such as morels) are also known to do this. Grasses grow wherever they can find a patch of sun, and mosses and lichens thrive on the damp ground and on the sides of tree trunks. In comparison with other biomes, however, the taiga has low biological diversity.

Jack pine cones and morels after fire in a boreal forest.

Coniferous trees are the dominant plants of the taiga biome. A very few species in four main genera are found: the evergreen spruce, fir and pine, and the deciduous larch. In North America, one or two species of fir and one or two species of spruce are dominant. Across Scandinavia and western Russia, the Scots pine is a common component of the taiga, while taiga of the Russian Far East and Mongolia is dominated by larch.

Fauna

Brown bear, Kamchatka peninsula. Brown bears are among the largest and most widespread taiga omnivores.

The boreal forest, or taiga, supports a relatively small range of animals due to the harshness of the climate. Canada's boreal forest includes 85 species of mammals, 130 species of fish, and an estimated 32,000 species of insects. Insects play a critical role as pollinators, decomposers, and as a part of the food web. Many nesting birds rely on them for food in the summer months. The cold winters and short summers make the taiga a challenging biome for reptiles and amphibians, which depend on environmental conditions to regulate their body temperatures, and there are only a few species in the boreal forest including red-sided garter snake, common European adder, blue-spotted salamander, northern two-lined salamander, Siberian salamander, wood frog, northern leopard frog, boreal chorus frog, American toad, and Canadian toad. Most hibernate underground in winter. Fish of the taiga must be able to withstand cold water conditions and be able to adapt to life under ice-covered water. Species in the taiga include Alaska blackfish, northern pike, walleye, longnose sucker, white sucker, various species of cisco, lake whitefish, round whitefish, pygmy whitefish, Arctic lamprey, various grayling species, brook trout (including sea-run brook trout in the Hudson Bay area), chum salmon, Siberian taimen, lenok and lake chub.

The taiga is home to a number of large herbivorous mammals, such as moose and reindeer/caribou. Some areas of the more southern closed boreal forest also have populations of other deer species such as the elk (wapiti) and roe deer. The largest animal in the taiga is the wood bison, found in northern Canada, Alaska and has been newly introduced into the Russian far-east. Small mammals of the Taiga biome include rodent species including beaver, squirrel, North American porcupine and vole, as well as a small number of lagomorph species such as snowshoe hare and mountain hare. These species have adapted to survive the harsh winters in their native ranges. Some larger mammals, such as bears, eat heartily during the summer in order to gain weight, and then go into hibernation during the winter.

Other animals have adapted layers of fur or feathers to insulate them from the cold. Predatory mammals of the taiga must be adapted to travel long distances in search of scattered prey or be able to supplement their diet with vegetation or other forms of food (such as raccoons). Mammalian predators of the taiga include Canada lynx, Eurasian lynx, stoat, Siberian weasel, least weasel, sable, American marten, North American river otter, European otter, American mink, wolverine, Asian badger, fisher, gray wolf, coyote, red fox, brown bear, American black bear, Asiatic black bear, polar bear (only small areas at the taiga – tundra ecotone) and Siberian tiger.

More than 300 species of birds have their nesting grounds in the taiga. Siberian thrush, white-throated sparrow, and black-throated green warbler migrate to this habitat to take advantage of the long summer days and abundance of insects found around the numerous bogs and lakes. Of the 300 species of birds that summer in the taiga only 30 stay for the winter. These are either carrion-feeding or large raptors that can take live mammal prey, including golden eagle, rough-legged buzzard (also known as the rough-legged hawk), and raven, or else seed-eating birds, including several species of grouse and crossbills.

Fire

The Funny River Fire in Alaska burned 193,597 acres (78,346 ha), mostly Black spruce taiga.

Fire has been one of the most important factors shaping the composition and development of boreal forest stands; it is the dominant stand-renewing disturbance through much of the Canadian boreal forest. The fire history that characterizes an ecosystem is its *fire regime*, which has 3 elements: (1) fire type and intensity (e.g., crown fires, severe surface fires, and light surface fires), (2) size of typical fires of significance, and (3) frequency or return intervals for specific land units. The average time within a fire regime to burn an area equivalent to the total area of an ecosystem is its *fire rotation* or *fire cycle*. However, as Heinselman noted, each physiographic site tends to have its own return interval, so that some areas are skipped for long periods, while others might burn two-times or more often during a nominal fire rotation.

The dominant fire regime in the boreal forest is high-intensity crown fires or severe surface fires of very large size, often more than 10,000 ha, and sometimes more than 400,000 ha. Such fires kill entire stands. Fire rotations in the drier regions of western Canada and Alaska average 50–100 years, shorter than in the moister climates of eastern Canada, where they may average 200 years or more. Fire cycles also tend to be long near the tree line in the subarctic spruce-lichen woodlands. The longest cycles, possibly 300 years, probably occur in the western boreal in floodplain white spruce.

Amiro calculated the mean fire cycle for the period 1980 to 1999 in the Canadian boreal forest (including taiga) at 126 years. Increased fire activity has been predicted for western Canada, but parts of eastern Canada may experience less fire in future because of greater precipitation in a warmer climate.

The Shanta Creek Fire began in a taiga area that had not had a major fire in over 130 years, and so was allowed to burn unchecked until it began to threaten populated areas.

The mature boreal forest pattern in the south shows balsam fir dominant on well-drained sites in eastern Canada changing centrally and westward to a prominence of white spruce, with black spruce and tamarack forming the forests on peats, and with jack pine usually present on dry sites except in the extreme east, where it is absent. The effects of fires are inextricably woven into the patterns of vegetation on the landscape, which in the east favour black spruce, paper birch, and jack pine over balsam fir, and in the west give the advantage to aspen, jack pine, black spruce, and birch over white spruce. Many investigators have reported the ubiquity of charcoal under the forest floor and in the upper soil profile. Charcoal in soils provided Bryson with clues about the forest history of an area 280 km north of the then current tree line at Ennadai Lake, District Keewatin, Northwest Territories.

Two lines of evidence support the thesis that fire has always been an integral factor in the boreal forest: (1) direct, eye-witness accounts and forest-fire statistics, and (2) indirect, circumstantial evidence based on the effects of fire, as well as on persisting indicators. The patchwork mosaic of forest stands in the boreal forest, typically with abrupt, irregular boundaries circumscribing homogenous stands, is indirect but compelling testimony to the role of fire in shaping the forest. The fact is that most boreal forest stands are less than 100 years old, and only in the rather few areas that have escaped burning are there stands of white spruce older than 250 years. The prevalence of fire-adaptive morphologic and reproductive characteristics of many boreal plant species is further evidence pointing to a long and intimate association with fire. Seven of the ten most common trees in the boreal forest—jack pine, lodgepole pine, aspen, balsam poplar (*Populus balsamifera*), paper birch, tamarack, black spruce—can be classed as pioneers in their adaptations for rapid invasion of open areas. White spruce shows some pioneering abilities, too, but is less able than black spruce and the pines to disperse seed at all seasons. Only balsam fir and alpine fir seem to be poorly adapted to reproduce after fire, as their cones disintegrate at maturity, leaving no seed in the crowns.

The oldest forests in the northwest boreal region, some older than 300 years, are of white spruce occurring as pure stands on moist floodplains. Here, the frequency of fire is much less than on adjacent uplands dominated by pine, black spruce and aspen. In contrast, in the Cordilleran region, fire is most frequent in the valley bottoms, decreasing upward, as shown by a mosaic of young pioneer pine and broadleaf stands below, and older spruce–fir on the slopes above. Without fire, the boreal forest would become more and more homogeneous, with the long-lived white spruce gradually replacing pine, aspen, balsam poplar, and birch, and perhaps even black spruce, except on the peatlands.

Threats

Human Activities

Plesetsk Cosmodrome is situated in the taiga.

Large areas of Siberia's taiga have been harvested for lumber since the collapse of the Soviet Union. Previously, the forest was protected by the restrictions of the Soviet Forest Ministry, but with the collapse of the Union, the restrictions regarding trade with Western nations have vanished. Trees are easy to harvest and sell well, so loggers have begun harvesting Russian taiga evergreen trees for sale to nations previously forbidden by Soviet law.

In Canada, eight percent of the taiga is protected from development, the provincial government allows forest management to occur on Crown land under rigorous constraints.

The main forestry practice in the boreal forest of Canada is clearcutting, which involves cutting down most of the trees in a given area, then replanting the forest as a monocrop (one species of tree) the following season.

Some of the products from logged boreal forests include toilet paper, copy paper, newsprint, and lumber. More than 90% of boreal forest products from Canada are exported for consumption and processing in the United States. Some of the larger cities situated in this biome are Murmansk, Arkhangelsk, Yakutsk, Anchorage, Yellowknife, Tromsø, Luleå, and Oulu.

Most companies that harvest in Canadian forests are certified by an independent third party agency such as the Forest Stewardship Council (FSC), Sustainable Forests Initiative (SFI), or the Canadian Standards Association (CSA). While the certification process differs between these groups, all of them include forest stewardship, respect for aboriginal peoples, compliance with local, provincial or national environmental laws, forest worker safety, education and training, and other environmental, business, and social requirements. The prompt renewal of all harvest sites by planting or natural renewal is also required.

Climate Change

Seney National Wildlife Refuge.

During the last quarter of the twentieth century, the zone of latitude occupied by the boreal forest experienced some of the greatest temperature increases on Earth. Winter temperatures have increased more than summer temperatures. The number of days with extremely cold temperatures (e.g., −20 to −40 °C (−4 to −40 °F) has decreased irregularly but systematically in nearly all the boreal region, allowing better survival for tree-damaging insects. In summer, the daily low temperature has increased more than the daily high temperature. In Fairbanks, Alaska, the length of the frost-free season has increased from 60–90 days in the early twentieth century to about 120 days a century later. Summer warming has been shown to increase water stress and reduce tree growth in dry areas of the southern boreal forest in central Alaska, western Canada and portions of far eastern Russia. Precipitation is relatively abundant in Scandinavia, Finland, northwest Russia and eastern Canada, where a longer growth season (i.e. the period when sap flow is not impeded by frozen water) accelerate tree growth. As a consequence of this warming trend, the warmer parts of the boreal forests are susceptible to replacement by grassland, parkland or temperate forest.

In Siberia, the taiga is converting from predominantly needle-shedding larch trees to evergreen conifers in response to a warming climate. This is likely to further accelerate warming, as the evergreen trees will absorb more of the sun's rays. Given the vast size of the area, such a change has the potential to affect areas well outside of the region.

In much of the boreal forest in Alaska, the growth of white spruce trees are stunted by unusually warm summers, while trees on some of the coldest fringes of the forest are experiencing faster growth than previously. Lack of moisture in the warmer summers are also stressing the birch trees of central Alaska.

Insects

Recent years have seen outbreaks of insect pests in forest-destroying plagues: the spruce-bark beetle (*Dendroctonus rufipennis*) in Yukon and Alaska; the mountain pine beetle in British Columbia; the aspen-leaf miner; the larch sawfly; the spruce budworm (*Choristoneura fumiferana*); the spruce coneworm.

Pollution

The effect of sulphur dioxide on woody boreal forest species was investigated by Addison et al., who exposed plants growing on native soils and tailings to 15.2 µmol/m³ (0.34 ppm) of SO_2 on CO_2 assimilation rate (NAR). The Canadian maximum acceptable limit for atmospheric SO_2 is 0.34 ppm. Fumigation with SO_2 significantly reduced NAR in all species and produced visible symptoms of injury in 2–20 days. The decrease in NAR of deciduous species (trembling aspen [*Populus tremuloides*], willow [*Salix*], green alder [*Alnus viridis*], and white birch [*Betula papyrifera*]) was significantly more rapid than of conifers (white spruce, black spruce [*Picea mariana*], and jack pine [*Pinus banksiana*]) or an evergreen angiosperm (Labrador tea) growing on a fertilized Brunisol. These metabolic and visible injury responses seemed to be related to the differences in S uptake owing in part to higher gas exchange rates for deciduous species than for conifers. Conifers growing in oil sands tailings responded to SO_2 with a significantly more rapid decrease in NAR compared with those growing in the Brunisol, perhaps because of predisposing toxic material in the tailings. However, sulphur uptake and visible symptom development did not differ between conifers growing on the 2 substrates.

Acidification of precipitation by anthropogenic, acid-forming emissions has been associated with damage to vegetation and reduced forest productivity, but 2-year-old white spruce that were subjected to simulated acid rain (at pH 4.6, 3.6, and 2.6) applied weekly for 7 weeks incurred no statistically significant (P 0.05) reduction in growth during the experiment compared with the background control (pH 5.6). However, symptoms of injury were observed in all treatments, the number of plants and the number of needles affected increased with increasing rain acidity and with time. Scherbatskoy and Klein found no significant effect of chlorophyll concentration in white spruce at pH 4.3 and 2.8, but Abouguendia and Baschak found a significant reduction in white spruce at pH 2.6, while the foliar sulphur content significantly greater at pH 2.6 than any of the other treatments.

Natural Disturbance

One of the biggest areas of research and a topic still full of unsolved questions is the recurring disturbance of fire and the role it plays in propagating the lichen woodland. The phenomenon of wildfire by lightning strike is the primary determinant of understory vegetation and because of this, it is considered to be the predominant force behind community and ecosystem properties in the lichen woodland. The significance of fire is clearly evident when one considers that understory vegetation influences tree seedling germination in the short term and decomposition of biomass and nutrient availability in the long term. The recurrent cycle of large, damaging fire occurs approximately every 70 to 100 years. Understanding the dynamics of this ecosystem is entangled with discovering the successional paths that the vegetation exhibits after a fire. Trees, shrubs, and lichens all recover from fire-induced damage through vegetative reproduction as well as invasion by propagules. Seeds that have fallen and become buried provide little help in re-establishment of a species. The reappearance of lichens is reasoned to occur because of varying conditions and light/nutrient availability in each different microstate. Several different studies have been done that have led to the formation of the theory that post-fire development can be propagated by any of four pathways: self replacement, species-dominance relay, species replacement, or gap-phase self replacement. Self replacement is simply the re-establishment of the pre-fire dominant species. Species-dominance relay is a sequential attempt of tree species to establish dominance in the canopy. Species replacement is when fires occur in sufficient frequency to interrupt species dominance relay. Gap-Phase Self-Replacement is the least common and so far has only been documented in Western Canada. It is a self replacement of the surviving species into the canopy gaps after a fire kills another species. The particular pathway taken after a fire disturbance depends on how the landscape is able to support trees as well as fire frequency. Fire frequency has a large role in shaping the original inception of the lower forest line of the lichen woodland taiga.

It has been hypothesized by Serge Payette that the spruce-moss forest ecosystem was changed into the lichen woodland biome due to the initiation of two compounded strong disturbances: large fire and the appearance and attack of the spruce budworm. The spruce budworm is a deadly insect to the spruce populations in the southern regions of the taiga. J.P. Jasinski confirmed this theory five years later stating "Their [lichen woodlands] persistence, along with their previous moss forest histories and current occurrence adjacent to closed moss forests, indicate that they are an alternative stable state to the spruce–moss forests".

Freshwater Swamp Forest

A freshwater swamp forest is defined as a forest that is full (inundated) of mineral-rich water (freshwater) either as a permanent, irregular or a seasonal condition. These forests are mainly found on the lower parts of a river or freshwater lakes in many different regions having a wide range of climate types. These include the northern areas which have a subarctic climate, tropical, subtropical, and temperate regions. They are different from peat swamp forests mainly because they do not have a thick layer of peat which is present in peat swamp forests due to lack of decomposition.

Freshwater swamp forests have multiple sources of water which include rivers, rain, and water from the ground. The waters' color in freshwater swamp forests can also differ depending on the number of plant materials found in the soil and water.

Freshwater swamp forests are usually endemic to the areas of Southeast Asia, South America, Africa and the Amazon where the most significant part of these forests occur. In Southeast Asia, the woods are found closer to large rivers like the Mekong, Irrawaddy, and Chao Phraya. Others, however, can be located near small streams.

These types of forests are mostly limited to alluvial soils carried by floods to the plains prevalent on a mangroves' forest landward side or areas with a shallow water table. Freshwater swamp forest, even though they are native to regions with a wet climate, they are also present in other areas like East Java and New Guinea which are dry seasonally.

Features

The aquatic flora present in these forests is primarily dependent on the nutrients in the water. Compared to forests on dry land, these swamp forests have few varieties of plants. Due to this, they are mostly full of one type or just a few species of trees. Thin peat, however, may be found on these forests contingent on the amount of water present in the forest due to the low rate of decomposition. This limited decomposition is caused by the presence of high amounts of phenolic compounds found in leaves which is sometimes thrice the amount in temperate forests. This means that decay occurs in anoxic conditions which in turn produces humid acid thus reducing the water pH. As a result, it makes trees acquires specific features like pneumatophores due to anaerobic conditions also caused by the frequent flooding, buttress roots, and lenticellate barks for stability and gas distribution respectively.

In fauna, freshwater swamp forests are just as diverse as dryland forests even though the research in this area is insufficient. They are known to be homes to numerous infrequent, endangered, and specialized species with 33% of birds and 45% of mammals in these forests having a status of either threatened or endangered on the IUCN Red List.

Threats

These forests are areas of global importance. They regulate the floods, maintain water quality, and conserve wetlands which include both flora and fauna. Despite their importance in preserving biodiversity, these forests are under threats. In many places, they face encroachment from people who seek to exploit their rich soils for agriculture. Their importance is also not well known which means that they are not being conserved as well as they should be. This can be seen in the Sumatran freshwater swamp forests and the Niger Delta freshwater swamp ecosystem.

References

- Green: expert-eyes.org, Retrieved 17 July, 2020
- Tropical-evergreen-rainforest-biome-location-and-climate-geography, forest-2922: geography-notes.com, Retrieved 06 August, 2020
- Temperate-coniferous-forest, biomes: worldwildlife.org, Retrieved 16 April, 2020
- Malhi, Yadvinder; Phillips, Oliver, eds. (2005). Tropical Forests and Global Atmospheric Change. Oxford, New York: Oxford University Press. doi:10.1093/acprof:oso/9780198567066.001.0001. ISBN 9780198567066. OCLC 77178196
- Ruckstuhl, K. E.; Johnson, E. A.; Miyanishi, K. (July 2008). "Boreal forest and global change". Philos. Trans. R. Soc. Lond. B Biol. Sci. 363 (1501): 2245–49. doi:10.1098/rstb.2007.2196. PMC 2387060. PMID 18006417
- What-is-a-freshwater-swamp-forest: worldatlas.com, Retrieved 25 March, 2020

3
Sustainable Forest Management

Forest management deals with the economic, social, technical, environmental and ecological aspects of forest. Sustainable forest management maintains balance between all these aspects by following the principles of sustainable management. This chapter has been carefully written to provide an easy understanding of sustainable forest management.

Forest Management

Forest Management are the various activities involved in regulating forest lands, describing them (forest valuation), taking inventory, studying forests, and working out long-term plans for the management of forest lands. Forest management establishes the main principles for the most efficient use and regeneration of forest resources, sets the age at which trees may be cut, calculates the amount of wood to be felled (optimum rate for annual cutting of forest), and establishes the methods and extent of improvement cuttings, reforestation, and other work. Special attention is paid to developing scientifically sound recommendations for the most rational exploitation of forest lands, increasing the productivity of forests, and expanding their uses. Efforts to plan and develop forest lands are being expanded. There will be further improvements in the comprehensive use of forests.

Forest management planning is one of the primary tools used to ensure that the country's publicly owned forests remain healthy and vibrant and are managed sustainably.

Key in this approach is the forest management plan that every forest company must, by law, draw up and have approved by government before any harvesting can begin on public lands.

A forest management plan, usually covering a period of several decades, must:

- Outline a strategic vision and a commitment to protect multiple forest values in the area under management.
- Assess the current state of the forest in the area the plan applies to.
- Detail the desired future state of forest values in the area.
- Set out the management objectives.
- Describe the harvesting, regeneration and other activities to be carried to achieve the stated objectives.

Forest management planning ensures that operations support sustainable forestry. Development of management plans follows a strict process which, although varying by jurisdiction, generally involves receiving input from industry, government agencies, the public and other stakeholders. The aim is to ensure that forestry operations and related activities are carried out in ways that support the sustainable management of all forest resources for generations to come.

In addition to forest management plans, more detailed tactical plans (to guide road building and harvest schedules, for example) are drawn up for specific operating areas and extend over a period of years. As well, annual operating plans identify exactly where and how harvesting operations and regeneration activities should occur in given years.

Even after a forest management plan is approved, government oversight continues. In their jurisdictions, provincial and territorial governments:

- Closely monitor forest companies to ensure they comply with the plans.
- Track the timber that is removed from the tenured lands.
- Ensure that the companies meet the regeneration standards after harvest.
- Audit compliance of the companies with regulations regarding the protection of soil, water and non-timber values and services.
- Review every company's forest plans every 5 or 10 years.

Forest companies are required to:

- Submit regular, formal reports on their activities.
- Update their plans in response to changing conditions and objectives.

Companies that fail to comply with forest management plans face penalties. These range from warnings and fines to the suspension of harvesting rights or seizure of timber.

Public Participation is Vital to Forest Management Planning

Public participation ensures that the planning process is transparent and gives people real influence in decision-making. Public participation processes vary from jurisdiction to jurisdiction but the parties generally include:

- Aboriginal peoples.
- Industry and environmental groups.
- Community representatives.

In many provinces, advisory groups must be involved in forest management planning for public lands and the plans must be made available to the public for review and comment.

Planning Process

This planning process includes careful identification of landowner objectives, inventory of resources, development and implementation of the management strategy to be used, and periodic re-evaluation of the implemented strategy.

Identify Landowner Objectives

The first step in developing a management plan is to identify landowner objectives—what products and amenities does the landowner wish to obtain from the land. It is imperative that these objectives be identified at the beginning because they determine what resources should be inventoried and define the goal of the management plan. Table lists some examples of forest landowner objectives.

Table: Examples of forest landowner objectives.

• Source of Income	• Speculation
• Tax Shelter	• Inheritance
• Product Yield	• Security
• Soil Stabilization	• Wildlife Habitat
• Aesthetics	• Recreation
• Preservation	• Increase Species Diversity
• Pride and Self-Satisfaction	• Exercise/Hobby

Some landowners may be interested in emphasizing only one management objective. They might, for example, be interested in maximizing the net financial return on investment through timber yield, or they might wish to develop the forest primarily as wildlife habitat. Management such as this, which emphasizes a single resource objective, is called dominant-use management. The forest can and will provide other products and amenities as it is managed (e.g., cuttings to enhance wildlife habitat will yield timber and income), but the management plan is developed to enhance or improve one (dominant) resource.

Most forest landowners, however, are interested in obtaining more than one product or amenity from their forests. They might, for example, desire income from timber harvest, wildlife habitat enhancement, and the maintenance of aesthetic quality. Forest management designed to enhance or produce more than one product or amenity is called multiple-use management. It is important to note that multiple-use management does not require that every acre of the forest be managed for every desired product or amenity, but rather that the forest as a whole be managed in such a way that it yields the desired mix. In the example given previously, some areas might be managed primarily for timber with aesthetic considerations, while on other acres specific practices are undertaken to enhance wildlife habitat.

Resource Inventory

Once ownership objectives have been defined, resources can be inventoried. Obviously, the tree/forest resources will be inventoried and such forest characteristics as tree species, condition, numbers, age, volume, value, growth, and basal area will be measured. Soil/site quality will be evaluated to determine what the site can produce. Depending on ownership objectives, other resources may be inventoried such as boundaries, wildlife, wildlife habitat, streams, trails, roads, campsites, vistas, and easements. Inventory data are analysed to determine what is present and what the forest site is capable of producing.

Develop and Implement the Management Strategy

Based on the inventory analysis, one or more management plans are then developed to achieve the ownership objectives. These forest management plans are based on and limited by what is biologically/ecologically possible on the area, what is economically and organizationally feasible, and what is socially and politically desirable.

The biological/ecological characteristics of the forest (e.g., tree species, soil type, topography, etc.) determine what is possible on the area, including such things as which tree species will grow, how fast they will grow, what wildlife will live on the area, etc. Based on the biological/ecological characteristics of the site, silvicultural practices can be prescribed to achieve ownership objectives. Silvicultural prescriptions are treatments designed to manipulate forested land such as various kinds of timber cuttings, tree plantings, prescribed burning, and the use of specific chemicals such as herbicides and fertilizers.

Economic/financial considerations may determine which activities are feasible. If economic/financial objectives are important to the landowner, then silvicultural activities undertaken must not only be biologically/ecologically possible, but also must contribute positively in the appropriate economic/financial analysis.

As an example, fertilization of most hardwood woodlands would result in an increase in total growth—it is ecologically possible and produces positive results. However, if the cost of the fertilizer was invested in a certificate of deposit, it would increase in value more than the forest will as a result of the fertilization. Therefore, if economic/financial return is an important ownership objective, fertilization is not an attractive silvicultural alternative.

On the other hand, there may be activities in which the landowner is willing to invest with little or no expected financial return because of the expected non-financial returns. These might include such things as the development of hiking trails, vistas, or wildlife habitat.

Similarly, limited financial or organizational resources may restrict management activities in a woodland. Obviously, if a landowner does not have the financial resources or labor to accomplish desired activities, they won't be done.

Finally, management activities are constrained by what is socially and politically desirable. Activities that violate the law are obviously unacceptable, and activities that upset neighbors are generally not prudent.

The careful development of a forest management plan in consultation with a professional forester is essential if landowners are to achieve their desired ownership objectives. The forest management plan is the blueprint of activities for caring for the forest. It is important to remember, too, that the management plan is not "cast in stone," but is an evolving plan that should be periodically reviewed and updated.

Purposes and Techniques of Forest Management

Multiple-use Concept

The forests of the world provide numerous amenities and ecosystem services in addition to being a source of wood products. The various public, industrial, and private owners of forestland may have quite different objectives for the forest resources they control. Industrial and private owners may be most interested in producing a harvestable product for a processing mill. However, they also may want other benefits, such as forage for grazing animals, watershed protection, recreational use, and wildlife habitat. On public lands the multiple-use land management concept has become the guiding principle for enlightened foresters. This is a complex ecological and sociological concept in contrast to the single-use principle of the past. The challenge, in the words of Gifford Pinchot, is to "ensure the greatest good for the most people over the long run." Thus timber production may have top priority in some areas, but in others, such as those near large population centres, recreational values may have high priority. Multiple use calls for exceptional skill on the part of forest managers.

Sustained Yield

Forest management originated in the desire of the large central European landowners to secure dependable income to maintain their castles and retinues of servants. Today forest management is still primarily economic in essence, because modern forest industries, mainly sawmilling and paper manufacture, can be efficient only on a continuous-operation basis.

Foresters think in long time scales, in line with the long life of their renewable crop. However, it is possible that a forest can be managed in such a way that a modest timber crop may be harvested indefinitely year after year if annual harvest and the losses due to fire, insects, diseases, and other destructive agents are counterbalanced by annual growth. This is the sustained-yield concept. An important element is the rotation, or age to which each crop can be grown before it is succeeded by the next one. Examples of short rotation periods in the subtropics are 7 years for leucaena for fuel wood, 10 years for eucalyptus, and 20 years for pine for pulpwood. Here a sustained yield could in theory be obtained simply by felling one-tenth of the eucalyptus forest each year and replanting it. Rotation periods for pulpwood in northern Europe and North America extend to 50 years. Softwood sawlogs often need 100 years to reach an economic size, while rotation periods for broad-leaved trees, such as oak and beech, in central Europe, may extend to two centuries. Over so long a growing spell only part of the lumber yield is obtained by the clear-cutting of a small fraction of the forest each year. The rest is secured by systematically thinning out the whole forest periodically.

Sustained-yield principles are likewise applied to minor forest produce. Turpentine and pitch are obtained by the systematic tapping of the lower trunk of certain subtropical pines. Successive cuts with a chisel-like tool every few days during a succession of summers eventually kill the trees. To ensure continued yields, crops of young pines are raised rotationally to replace those felled. A similar system is followed for the rubber tree (Hevea brasiliensis) that is grown in plantations.

Forest Products

The culture of trees in natural forests and plantations for the yield of lumber, pulp, chips, and specialty products is a principal management objective. In many parts of the world the harvest of wood for firewood and charcoal is the dominant use, and these products are often in short supply. Timber stands must be felled and regenerated in an orderly sequence to meet continuing industrial demands.

Natural Regeneration

In established forests the selective cutting of marketable timber, taking either one tree at a time (single-tree selection) or a number of trees in a cluster (group selection) and leaving gaps in which replacements can grow up from natural seedlings, can prove economical and also ensure the best possible use of available soil, light, and growing space. The best examples of single-tree-selection forests are found in Switzerland, on slopes where any clear felling could lead quickly to soil erosion and avalanches.

Alternative methods of natural regeneration deal with areas of land as units, rather than with single trees. One highly effective example is employed in the Douglas fir forests along the Pacific slope of Canada and the western United States. Logging by powerful yarding machines, using overhead cables, creates wedge-shaped gaps of cleared ground. The surrounding forest is left standing for many years in order to provide shelter and seed. Abundant seed is carried by wind on to the cleared land and gives rise, in a few years, to a full crop of seedling firs. After these have reached seed-bearing age, the areas previously left standing may be removed in their turn. Similar systems using a pattern of strips cut across the forest, or circular plots gradually extended until they meet and coalesce, are employed in France and Germany.

A silvicultural system employing practices of short rotation (five to 10 years) and intensive culture (fertilization, weed, and insect and disease control) with superior genotypes relies on coppice, or regeneration from sprouts arising from stumps of felled trees, as the method of regeneration of the new crop and is characterized by high productivity.

Artificial Regeneration

Artificial regeneration is accomplished by the planting of seedlings (the most common method) or by the direct planting of seeds. Direct seeding is reserved for remote or inaccessible areas where seedling planting is not cost-effective. A few tree species, such as poplars (Populus species) and willows (Salix species), are artificially reproduced from cuttings. Most forest planting in North America involves the conifers, especially the pines, spruces, and Douglas fir, because of the prospects of successful establishment and high financial yield. The amount of hardwood planted worldwide has increased from earlier periods, with major gains in tropical hardwoods (Eucalyptus species, Gmelina species) and high value temperate species.

Artificial regeneration offers greater opportunity than natural regeneration to modify the genetic constitution of stands. The most important decision made in artificial regeneration is the selection of the species used in each new stand. The species chosen should be adapted to the site. The most successful introductions are obtained by moving species to the same latitude and position on the continent that they occupied in their native habitat. For example, many conifers of the western coasts of North America have been successful at the same latitudes in Western Europe. The forest economy of many countries in the Southern Hemisphere is dependent on pines introduced from localities of comparable climate in the southern United States, California, and Mexico.

The variability of seed quantity and quality and the demand for superior genotypes has led to the creation of seed orchards, stands of trees selected for superior genetic characteristics, which are cultivated to produce large quantities of seed. Most kinds of seed can be stored in sealed containers in refrigerators at temperatures near freezing for several years without a significant loss in viability. For some species, a brief period of cold storage may be necessary for the seeds to germinate; this stratification treatment is needed to satisfy the dormancy requirement of some temperate-zone species.

Direct sowing of harvested seed in the forest or on open land is not a common practice because of forest seed-eaters (mice, squirrels, birds) and the problem of weed growth. Tree seedlings are therefore raised in forest nurseries, where effective protection is possible. These seedlings almost invariably come from seed, although vegetative propagation from rooted cuttings is a useful technique of perpetuating valuable strains of certain species. Seedlings grown in raised seedbeds are removed from the nursery soil when large enough and are bare-rooted when planted in the field. Seedlings grown in individual containers have an intact root system encapsulated in a soil plug for planting. In either case, the system can be highly mechanized. To enhance seedling quality, the seedbeds or container media are inoculated with specific microorganisms that form symbiotic relationships with the seedlings. These microorganisms include certain fungi, which form mycorrhizae with the roots and improve nutrient and water uptake, and nitrogen-fixing bacteria such as Rhizobium species and Frankia species, which contribute nutrients. Selective herbicides, insecticides, and fungicides are applied before or after seedling emergence to keep the developing seedlings free of weeds, insects, and disease.

Many tree seedlings are suitable for field planting after a few months in a containerized seedling nursery or after one to two years in a seedbed. Slow-growing species are transplanted by hand or machine during the dormant season to transplant beds where they are root-pruned and fertilized to stimulate top growth and the development of a bushy root system, characteristics essential for survival in field planting. The mechanized operation is highly efficient. One machine and four workers can transplant 30,000–40,000 seedlings each working day. Weeds are controlled during the transplant stage by chemical herbicides that inhibit weed seed germination or growth or by mechanical harrows drawn between the rows.

In preparation for field planting, dormant nursery-grown seedlings are undercut with a sharpened steel blade and removed from the bed by hand or by a mechanized vibrating lifter and conveyor belt system. Roots of seedlings lifted in autumn are packed with moistened sphagnum moss, and the bundles are stored in refrigerated coolers. Alternatively, seedlings may be placed in a trench or heeling-in bed, and covered with soil and mulch until spring. At the time of lifting, seedlings should be culled to eliminate those that will not survive after planting—i.e., seedlings infested with insects or disease, badly damaged in lifting and handling, having distinctly poor root systems, or falling below minimum size standards. It is imperative that the seedlings be kept cool and the root systems moist in all phases of the lifting, storage, transport, and planting processes.

Container-grown seedlings are culled in a manner similar to the bare-rooted stock and in most cases are shipped in the containers in which they were produced. The container method, which has traditionally been used in the tropics or in locations that are hot and dry, has become the principal method of seedling production in Canada, Scandinavia, and portions of continental Europe, Japan and China.

Planting tree seedlings is one of the most costly investments in the production of a forest crop. The success of a whole rotation is often determined by the soundness of decisions made about planting. These decisions concern the selection of the planting stock, the density of the planting, the use of mixed plantings, the season of planting, preparation of the site prior to planting, and even the method of planting. In temperate climates planting is generally conducted from late winter to late spring, but the use of container-grown seedlings extends the planting season into the early summer and includes a period in early autumn.

On level ground, machine planting is preferred over hand planting. A planting machine forms a groove in the soil in which seedlings are placed at specified intervals; a set of blades then cuts into the soil around the planted seedling, and a set of packing wheels firms the soil around it. A planting machine pulled behind a single tractor on prepared level ground can set 8,000–10,000 seedlings per day. On steep slopes, broken or rocky ground, or amid tree stumps and tops, planting is done by hand. The planter uses a spade, planting bar, or mattock (or a variation of one of these) to cut a notch, or dig a pit, into which the seedling roots are inserted. Soil is then replaced and stamped firmly around the base of the seedling.

During the following growing season, and possibly two to three years thereafter, weed control may be essential for the survival and early growth of the planted seedling. Weeds may be removed by hand with a sharp tool or hoe or by other mechanical means such as mowing or cultivating between the planted rows. Herbicides may offer a more effective and efficient means of weed control. While care must be exercised to shield the tree from many chemicals, compounds are available that kill unwanted vegetation but do not harm the tree seedling. In some regions the lower branches of conifers and certain highly valued hardwoods are pruned from saplings and young trees to improve the quality and value of the main stem and improve access into the plantation. Otherwise, the artificially established plantation needs, and receives, no more attention than does the naturally regenerated crop.

Until the 20th century foresters usually accepted the land much as they found it. Their reaction to infertile soil was to plant aggressive species of trees, regardless of their potential market value, and to accept lower returns in plant production. Development of modern machines and a growing understanding of plant nutrition and soil chemistry now enable foresters to improve sites much as a farmer does and thereby to increase output substantially. Mechanical draining, using tractor-drawn plows to create deep open drains and so aerate the soil, is now used on the peaty swamps of Europe, especially in Finland. On the hard heathlands of the United Kingdom, areas of new afforestation land have been broken up with sturdy plows designed to turn over firmly compacted soil layers. Plowing facilitates penetration of air, water, and tree roots; checks weed growth; and lessens fire hazard, though it may also disrupt beneficial microorganisms in the soil and can encourage the germination of weeds from the soil seed bank. So far it has usually been confined to strips for each row of trees, but full plowing as done on a farm promises further advantages.

In poorly drained or coastal areas, sites are prepared by a bedding plow, which creates an alternative ridge and valley surface that improves soil drainage, aeration, and nutrient availability. Subsequent to bedding, seedlings are planted on the ridge or bed. Because forest crops are rarely irrigated (returns are too low for the capital cost invested), forest plantings on droughty sites require a careful selection of the species and the time for planting and an effective weed control program.

The fundamental relationship between mineral nutrition and growth is the same for trees as for other plants. An understanding of forest tree nutrition requires recognition of factors distinctive to forests:

- The nutrient demands of the plantation vary from season to season and with the developmental stage of the stand. During the life of a forest tree crop, large quantities of nutrients are returned to the soil in organic matter, which is, in turn, mineralized and made available for reuse by the same or the following crop.
- Retranslocation of absorbed nutrients is highly developed in trees; i.e., nutrients in leaves move back into stems prior to fall leaf drop and then move into new leaves in the spring.
- Except for the first year after planting, trees start the growing season with a developed framework for photosynthesis and an established root system for nutrient and water uptake.
- The use of soil resources such as water and nutrients by trees may often be strongly influenced by mechanisms involved in adaptations for survival from one season to another, rather than in growth.

Judicious management of nutrition ensures not only increased productivity of existing forests but also sustained productivity over many rotations. In southern Australia, for example, declines in yield of 25–30 per cent in second rotation Monterey pine (Pinus radiata) plantations have been corrected by a number of means, including intensive silviculture (site preparation, weed control, and fertilization) during the early stages, retention and management of forest debris (leaves, branches, etc.) to conserve nutrients, and intercropping with annual legumes, which supply nitrogen and other nutrients.

Range and Forage

Important among the broad spectrum of forest resources are the understory plants that can provide forage for grazing animals, both domestic and wild. Grazing livestock are useful to the forest manager. Dense old-growth forest or vigorous second-growth stands with closed canopies generally have sparse, low-quality forage. Large forest management units, however, generally contain extensive logged or burned areas where understory forage plants temporarily dominate the site. These areas are transitory since the tree canopies close in 10 to 20 years, but they can provide good forage until canopy closure. Cutting cycles in the managed forest and even wildfires provide a continuing grazing resource that shifts from one location to another. In addition, open meadows occurring in valley bottoms, open forests on shallow soils, and grassland balds on windswept ridge tops greatly enrich the grazing potential of the forest. Grazing fees offset the long-term investments that must be carried in renewing the forest.

Hardwood forests are more susceptible than coniferous forests to grazing damage. The current year's growth on broad-leaved trees provides palatable forage during most seasons of the year, whereas coniferous needles are much less palatable. Uncontrolled livestock-grazing in some parts of the world has been particularly devastating to forests and is a serious problem.

In agroforestry, a practice known as silvopasture, or dehesa, specifically seeks to combine trees with forage (pasture) and livestock production. The components are structurally and functionally combined and actively managed to optimize the positive biophysical interactions between them. This form of agroforestry is a practical and low-cost means of implementing integrated land management (which seeks to reduce human impacts on land), and promotes long-term, sustainable, and renewable forest management, especially for small-scale producers.

Recreation and Wildlife

From the earliest times human beings have looked to the forests for recreation. Today, recreation in forests assumes ever-growing importance with the growth of cities whose inhabitants need a change of scene, fresh air, and freedom to wander, as a relief to the stresses of industrial and commercial life. Imaginative planning is essential to ensure that people actually find what they are seeking without damage to the forest environment or conflict with the pleasures of others. The most popular outdoor recreation activities utilize forestland and include hunting and fishing, picnicking and camping, hiking, mountain climbing, driving for pleasure, boating and other water sports, winter sports, photography, and nature study. The challenge is to balance the varied demands for recreational use with the other forest uses.

For many recreationists the main attraction of the woods is the abundance of animal and plant life. The forest manager must attempt to satisfy the diverse needs of hunters and sportsmen, outdoorsmen, and preservationists. This requires a broad expertise drawing on principles from the social sciences, natural history, wildlife management, conservation, landscape design, law, and public administration, among other disciplines.

Recreation management includes visitor management as well as resource management. Reasonably accurate assessments of the type and amount of use that areas receive are important to allow for efficient allocation of budgets and employee time and to ensure that the degree of use does not cause excessive impacts on resources and thus destroy the recreational value of the site. Skilful location of roads, picnic points, parking lots, and campgrounds ensures that the great majority of visitors congregate in relatively small portions of a large forest. Visitor management for some situations can be aided by use of computer-generated simulation models.

Some types of recreation require intensive management and special amenities. Vehicular camping facilities, for example, are designed for intensive use by large numbers of people and typically provide electrical hook-ups, toilets, showers, picnic tables, fireplaces, garbage receptacles, directional and interpretive signs, and play areas. These features must be durable and easily maintained. Downhill ski areas are most popular when well equipped with various runs, lifts, restaurants, and lodges. Other types of recreation, such as trail hiking and cross-country skiing, demand larger tracts of land but fewer improvements. Wilderness areas afford the personal challenge and serenity of backpacking, tent camping, and canoeing.

The interpretation of what visitors see in the forests has become a growing activity of most forest services. Nature trails, guidebooks, signposts, interpretive museums, and information stations assist visitors who come to learn as well as to enjoy.

Forests contain natural habitats for a wide range of wildlife, from the elks, wolves, lynxes, and bears of northern coniferous forests to the antelopes, giraffes, elephants, monkeys, lions, and tigers of tropical savannahs and jungles. Certain birds, such as pheasants, wood grouse, and quail, have high sporting value, while others are cherished for attractive song, appearance, or rarity. Many endangered species depend on forest habitats that are carefully protected by national and international laws.

Forest managers must attend to the interrelated, and sometimes directly opposed, wildlife interests of hunters, conservationists, and farmers. Obviously the same animal can present a different aspect to each group. A Bengal tiger, for example, provides a biologist with a classic example of a carnivorous beast living in harmony with a jungle environment and restraining its main prey, deer, from undue increase in numbers. But to a village peasant it is a menace to cows and goats and a threat to human safety, while a game hunter regards it as a magnificent quarry demanding skill. The needs of the forest itself require the numbers of grazing and browsing animals to be kept to a tolerable level. Otherwise renewal of tree crops becomes impossiblc.

Virtually every change that occurs in a forest benefits some wildlife species and harms others, though plantations of monocultures are generally poor habitats. Some wildlife species require a diversity of conditions—one type for feeding, another for nesting, and yet another for cover. Some have very specific requirements essential to their existence, whereas others have a broad range of tolerance. In any case, the life history characteristics of the species must be known in order for the resource manager to plan and implement practices necessary for the well-being of the species. Sometimes the best management involves increasing the forest edge habitat, frequented by many kinds of wildlife. Forest edge improvement may be integrated with timber harvesting and the construction of fire lanes and logging roads. Because food and cover for wildlife are often more plentiful in the early stages of forest development, retardation of succession by prescribed burning may be beneficial to wildlife. Food crops may be planted in certain areas to improve the wildlife-carrying capacity. Adjustments are often made by foresters in cutting procedures, rotation age, regeneration methods, and other practices to accommodate the food and cover needs of wildlife and fish. Certain areas may be managed exclusively for wildlife, particularly in situations where habitat for endangered species must be protected.

In virtually every country the sporting aspect of woodland wildlife management is controlled, to some degree, by general game laws, which also apply outside the forests. These laws prescribe licenses for firearms and the taking of specified birds and beasts; they usually lay down closed seasons during which certain game may not be shot and also set limits on the killing or capture of rare species. In the United States a peculiar situation exists whereby the game legislation of the separate states applies unchanged over most publicly owned forests. In other countries the forest managers are in a stronger position, since local game laws are adjusted to their particular requirements.

Watershed Management and Erosion Control

Not only is the presence of water in soils essential to the growth of forests, but improved water yield and quality are becoming increasingly important management objectives on many forested lands. Forests and their associated soils and litter layers are excellent filters as well as sponges, and water that passes through this system is relatively pure. Forest disturbances of various kinds can speed up the movement of water from the system and, in effect, reduce the filtering action. While disturbances are inevitable, in most instances they need not contribute to poor water quality.

In mountainous territory the value of forests for watershed and erosion protection commonly exceeds their value as sources of lumber or places of recreation. The classic example is found in Switzerland and the neighbouring Alpine regions where the existence of pastoral settlements in the valley is wholly dependent on the maintenance of continuous forest cover on the foothills of the great peaks. This is combined skilfully with limited lumbering and widespread recreational use by tourists.

The guiding principle of management where erosion threatens is therefore the maintenance of continual cover. Ideally, this is achieved by single-stem harvesting; only one tree is felled at any one point, and the small gap so created is soon closed by the outward growth of its neighbours.

The progress of water, from the time of precipitation until it is returned to the atmosphere and is again ready to be precipitated, is called the hydrologic cycle. The properties of the soil plant system provide mechanisms that regulate interception, flow, and storage of water in the cycle. The water that moves downward into the soil, or infiltrates, is the difference between precipitation and the losses due to canopy interception, forest floor interception, and runoff. The amount of water stored in the soil is largely dependent on the physical properties of the soil, its depth, and the amount of water lost due to evaporation from the soil surface and transpiration from plants (evapotranspiration). Transpiration is the water absorbed by plant roots that is subsequently evaporated from their leaf surfaces. Deep forest soils have a high water-storage capacity. Unless they are very porous and drain freely, they have a water table below which the subsoil is saturated. The depth of the water table varies seasonally and is higher during periods of low evapotranspiration. Removal of the forest canopy in wet areas also raises the water table. Most tree roots need air to survive and cannot exploit soil below the water table. The drainage of land having a high water table usually increases the productivity of the forest.

When incoming precipitation exceeds the soil's water-storage capacity, the excess water flows from the soil and can be measured as stream flow. The water yield of a forest is a measure of the balance between incoming precipitation and outflow of water as streamflow. The amount of increase in water yield depends on annual precipitation as well as the type and amount of overstory vegetation removed. As forests regrow following cutting, increases in streamflow decline as a result of increased transpirational losses. Streamflow declines are greater in areas that are restocked with conifers than in those restocked with hardwoods. This results from greater transpiration losses during the winter months from coniferous species.

Despite the uncertain balance of water gain and loss, forests offer the most desirable cover for water management strategies. Water yields are gradual, reliable, and uniform, as contrasted to the rapid flows of short duration characteristic of sparsely vegetated land. Unforested land sheds water swiftly, causing sudden rises in the rivers below. Over a large river system, such as that of the Mississippi, forests are a definite advantage since they lessen the risk of floods. They also provide conditions more favourable to fishing and navigation than does unforested land. All natural streams contain varying amounts of dissolved and suspended matter, although streams issuing from undisturbed watersheds are ordinarily of high quality. Waters from forested areas are not only low in foreign substances, but they also are relatively high in oxygen and low in temperature. Nonetheless, some deterioration of stream quality can be noted during and immediately after clear-cut harvesting, even under the best logging conditions. The potential for water-quality degradation following timber harvest may involve turbidity (suspended solids) as well as increases in temperature and nutrient content. Sediment arising from logging roads is the major water-quality problem related to forest activities in many areas.

The belief that forests increase rainfall has not been substantiated by scientific inquiry. Local effects can, however, prove substantial, particularly in semiarid regions where every millimetre of rain counts. The air above a forest, as contrasted with grassland, remains relatively cool and humid on hot days, so that showers are more frequent. Fog belts, such as those found along the Pacific seaboard of North America and around the peaks of the Canary Islands, give significant water yields through the interception of water vapour by tree foliage. The vapour condenses and falls in a process described as fog drip.

Fire Prevention and Control

A forest fire is unenclosed and freely spreading combustion that consumes the natural fuels of a forest—i.e., duff, grass, weeds, brush, and trees. Forest fires occur in three principal forms, the distinctions depending essentially on their mode of spread and their position in relation to the ground surface. Surface fires burn surface litter, other loose debris of the forest floor, and small vegetation; a surface fire may, and often does, burn taller vegetation and tree crowns as it progresses. Crown fires advance through the tops of trees or shrubs more or less independently of the surface fire and are the fastest spreading of all forest fires. Ground fires consume the organic material beneath the surface litter of the forest floor; ground fires are the least spectacular and the slowest-moving, but they are often the most destructive of all forest fires and also the most difficult to control.

A forest fire does a number of specific things. First, and perhaps most obviously, it consumes woody material. Second, the heat it creates may kill vegetation and animal life. In most fires, much more is killed, injured, or changed through heat than is consumed by fire. Third, it produces residual mineral products that may cause chemical effects, mostly in relation to the soil. The lethal temperatures for the living tissues of a tree (i.e., the phloem and cambium, which are located under the bark) begin at 49 °C (120 °F) if exposure is prolonged for one hour. At 64 °C (147 °F) death is almost instantaneous. The ignition temperature for woody material is approximately 343 °C (650 °F), with a flame temperature of 870–980 °C (1600–1800 °F).

Forest fires seldom occur in tropical rainforests or in the deciduous broad-leaved forests of the temperate zones. But all coniferous forests, and the evergreen broadleaf trees of hot, dry zones, frequently develop conditions ideally suited to the spread of fire through standing trees. For this, both the air and the fuel must be dry, and the fuel must form an open matrix through which air, smoke, and the gases arising from combustion can quickly pass. Hot, sunny days with low air humidity and steady or strong breezes favour rapid fire spread. In coniferous forests the resinous needles, both living and dead, and fallen branch wood make an ideal fuel bed. The leaves of evergreen broadleaf trees, such as hollies, madrone, evergreen oaks, and eucalyptus, are coated in inflammable wax and blaze fiercely even when green. Once started, fire may travel at speeds of up to 15 kilometres (10 miles) per hour downwind, spreading slowly outward in other directions, until the weather changes or the fuel runs out.

Well over 95 percent of all forest fires are caused by people, while lightning strikes are responsible for 1–2 percent. In some countries the setting of fires for clearing cropland is an integral technique of agriculture. In other areas forest fire prevention, including public education, hazard reduction, and law enforcement, consumes a considerable amount of time and money. The two basic steps in preventing forest fires are reducing risk and reducing hazard. Risk is the chance of a fire's starting as determined by the presence of activity of causal agents, most likely human beings. Hazard is reduced by compartmentalizing a forest with firebreaks (alleyways in which all vegetation is removed) and reducing the buildup of fuel (litter, branches, fallen trees, etc.) by controlled burning. In the United States the Forest Service devised a National Fire-Danger Rating System, which is the resultant of both constant and variable fire danger factors that affect the inception, spread, and difficulty of control of fires and the damage they cause.

Effective fire control begins with a field survey and map to identify the areas at risk, delineate them, and define and improve the barriers or firebreaks that may limit fire spread. Natural barriers include rivers, lakes, ridge tops, and tracts of bare land. Artificial barriers can be roads, railways, canals, and power-line tracks, but usually extra firebreaks must be cut to link these and provide wider gaps that fire cannot readily jump. Belts of land from 10 to 20 metres (33 to 66 feet) wide are cut clear of trees or left unplanted when a new forest is formed. Sometimes the soil is left bare and cultivated only at intervals to check invasion by weeds. Usually it is sown with an even crop of low perennial grasses or clovers and kept short by mowing or grazing. This checks soil erosion, provides an evergreen fireproof surface, and allows access on foot, by car, or in an emergency by fire-fighting trucks. Surfaced roads, serving also for lumber haulage and access for recreation, are of critical importance in fire fighting. Signposts are needed to guide fire crews unfamiliar with the woods and to mark water supplies and rendezvous points.

Detection is the first step in fire suppression. Many countries have organizations of trained professionals to detect and fight fires; others rely on volunteers or a combination of the two. Tower lookouts are the mainstay of nearly all detection systems, although the use of aircraft and satellites has modified this view in countries with an advanced fire control program. Fire surveillance is essential during seasons of high risk. Towers are set on hilltops where observers equipped with binoculars, maps, and a direction scale determine the compass direction of smoke and notify the fire control base via telephone or radio. If a fire can be seen from two or more towers, its precise position is quickly determined by mapping the intersection of cross bearings. Aircraft are used to detect fires and to carry out reconnaissance of known fires. Aerial surveillance has probably been most successful in detecting lightning-caused fires and is most often employed in areas of relatively low-value lands and inaccessible areas. An aircraft is essentially a moving fire tower, and the problems of detection that apply to a tower also apply to an aircraft; however, new developments in remote-control television, high-resolution photography, heat-sensing devices, film, and radar make fire detection by aircraft and satellite more efficient and location more accurate. Satellites provide a rapid means of collecting and communicating highly precise information in fire detection, location, and appraisal.

Once a fire has been detected, the next step is fire suppression. The first job is to stop or slow the rate of spread of the fire, and the second job is to put it out. The aim of suppression is to minimize damage at a reasonable cost. This does not necessarily mean the same thing as minimizing the area burned, but it is a major goal. Suppression is accomplished by breaking the "fire triangle" of fuel, temperature, and oxygen by robbing the fire of its fuel (by physically removing the combustible material or by making it less flammable through application of dirt, water, or chemicals); by reducing its temperature (through application of dirt, water, or chemicals and partial removal or separation of fuels); and by reducing the available oxygen (by smothering fuels with dirt, water, fog, or chemical substances).

The great majority of all forest fires are contained by professional fire fighters equipped with numerous hand tools (spades, beaters, axes, rakes, power saws, and backpack water pumps). Trained fire crews with light, hand equipment can be carried quickly to a fire by truck, delivered by helicopter, or even dropped by parachute. When necessary, large machines (bulldozers or plows) are used to clear openings, or firebreaks, which stop the spread of the fire. This requires clearing surface and sometimes aerial fuels from a strip of land and then digging down to mineral soil to stop a creeping or surface fire. A control line can also be established by directly extinguishing the fire along the edge or by making fuels non-flammable. In some cases a backfire may be deliberately set between the control line and the oncoming fire to burn out or reduce the fuel supply before the main fire, or head fire, reaches the control line.

Water is the most obvious, efficient, and universal fire extinguisher, but large-scale use of water in fire fighting is limited because it is usually in short supply and application methods are not adequate. For these reasons other materials have been tested for persistence and efficiency in putting out fires. Wetting agents change the physical characteristics of water to increase its penetrating and spreading abilities. Retardants, such as sodium calcium borate, reduce the flammability of wood and therefore its rate of burning. Foaming agents in powder or liquid form can greatly increase the mixture volume and thereby cool, moisten, and insulate the fuel.

Aircraft can quickly carry in water and other chemicals to be dropped or sprayed on the fire. A method developed on Canadian lakes is to fill the floats of a seaplane with water, which is done as it skims the lake on take-off, and to discharge this through nozzles over the fire.

The prescribed use of fire in forestland management has been approached with understandable reluctance by many foresters and wild land managers. Yet, fire has a place in the management of particular ecosystems. The decision to use fire is usually based on a balancing of pros and cons—i.e., damage, possible or expected, must be weighed against benefits. Under proper circumstances, prescribed burning can be used to prepare seedbeds for natural germination of most tree species, to control insect and disease infestations, to reduce weed competition, to reduce fire hazard, and to manipulate forest cover type.

Insect and Disease Control

Enormous numbers and varieties of insects, fungi, bacteria, and viruses occur in forests and are adapted to live on or around trees. Many of these are beneficial, and even the destructive ones are usually held in check by their natural enemies or an unfavourable environment. The normal population levels of pest organisms result in limited reduction in tree growth or the total destruction of only a small number of trees in the forest. The losses are generally accepted by foresters as unavoidable and are tolerated as long as the annual destruction does not seriously affect the net annual increase in wood production.

Every part of a growing tree—root, trunk, bark, leaf, flowers, and seeds—is potentially subject throughout every stage of its life to attack by some harmful insect or fungus. Insects actually destroy more standing timber than does any other agent. Bark beetles, including species of Dendroctonus and Ips, are among the most destructive insects. They bore into the tree and feed just below the bark, where they create tiny channels that disrupt the flow of food to the roots, often killing the tree. Diseases frequently retard growth of trees and are less of a factor in mortality. A particularly destructive disease is caused by fungi that decay the wood of trees. The heartrot fungi gain entrance through any wound resulting from fire scars, broken limbs, or anything else that damages the tree's protective tissue. Were it not for heartrot, a large number of conifers and broadleaved trees could be left to grow for many more years.

Asian longhorned beetle: Asian longhorned beetle (Anoplophora glabripennis) emerging from an exit hole in an infested tree.

Insect and disease organisms accidentally introduced to forests from other parts of the world, known as invasive species, often develop serious epidemic conditions because of the lack of any natural control. Because of rapid global transportation, insects and fungal spores can be spread easily throughout the world and arrive in a healthy condition. The seriousness of the situation cannot be overestimated, and the enforcement and improvement of plant quarantine laws is essential. Typically disasters have arisen where quarantine has failed or has been imposed too late. The American chestnut, Castanea dentata, has been virtually wiped out by the chestnut blight fungus, Endothia parasitica, which does little harm to related trees in its native China. Elms have suffered severely, both in Europe and in the United States, from the elm disease fungus, Ceratocystis ulmi, which was first detected in the Netherlands and is carried from tree to tree by flying beetles. Minute aphids, probably introduced on living plants from Asia, now make it impossible to raise commercial crops of two conifers once valued in Britain, namely, the white pine, Pinus strobus, from New England, and the European silver fir, Abies alba, native to Switzerland.

Generally the healthier the forest, the more resistant it is to widespread pest attack. Overmature, weak, wind-thrown, and lightning- or fire-killed trees have little or no defense against infestation and are a factor in the build-up of pest populations. Selective cutting of susceptible trees, thinning that accelerates growth, and other similar long-range forest management practices that stimulate vigorous tree growth are good methods for indirect control of insects and diseases. These practices reduce the host material and breeding grounds of pests that may spread to healthy trees. In regions with a high incidence of a known pest, foresters attempt to avoid serious trouble by planting only trees known to resist existing pests in the regions where the trees are grown. Many forest genetic programs have as a major goal the selection and breeding of trees with insect and disease resistances.

Occasionally the natural conditions that suppress the population of pest organisms change, and outbreaks in forests may reach epidemic proportions. Even-aged stands and plantations with trees of the same species and of uniform size and age often create perfect conditions for the rapid spread of insects and diseases. Even the more complex uneven-aged forests with their inherent check-and-balance systems can develop devastating populations of pests. At this point the forester must consider direct control measures.

Because effective direct control of insects and diseases of standing timber is generally expensive, it is employed only when the potential mortality or loss in growth is extreme. Routine monitoring of insects and diseases allows foresters to schedule timely harvests of infested trees and to limit the spread of the problem to uninfested trees or areas. These sanitation and salvage harvests, coupled with piling and burning the limbs and branches left after logging, reduce the material and conditions that allow pest populations to develop. Traps baited with sex-attractant chemicals, or pheromones, are a promising method to reduce breeding populations of certain insects. Application of insecticidal or fungicidal sprays from the ground or from low-flying aircraft offers a short-term measure to check sudden plagues of insects or outbreaks of fungal diseases. Action has most frequently been taken against exceptional outbreaks of defoliating caterpillars, including those of the gypsy moth in the United States, the nun moth in central Europe, and the pine looper moth in England. At the time of year when feeding caterpillars are most vulnerable, light aircraft fly across the forest on carefully planned courses, distributing pesticides.

A disadvantage of these blanket treatments by potent, broad-spectrum chemicals is that they also eliminate parasitic and predatory insects that serve as natural controls on the pest's numbers. They may also adversely affect birdlife and can contaminate waterways and affect aquatic invertebrates. In practice, large-scale chemical treatments of forests are infrequent and are restricted to a small proportion of the areas at risk. Generally, natural control through predatory organisms, which also cycle opportunistically in a slightly delayed sequence with the pest populations, combined with physical factors like cold winters, provides adequate checks. Biological control involving the release of predators or diseases of pests is promising in some situations.

Less spectacular preventive measures are commonly taken as routine steps in practical forestry to lessen anticipated losses. Nursery stock, easily reached and handled, may be grown in fumigated seedbeds and sprayed during production so that uninfested and vigorous seedlings are planted in the forests. Prompt removal of logs from the forest to distant processing mills transfers beetles that may emerge from beneath the bark to areas where they can do no harm. Stumps of freshly felled conifers can be easily and cheaply treated by brushing on a fungicide to check the white root-rot fungus, Fomes annosus. This is a serious agent of decay that spreads underground through root grafts after gaining entry via the exposed surface of a felled stump.

Urban Forestry

Urban forestry, which is the management of publicly and privately owned trees in and adjacent to urban areas, has emerged as an important branch of forestry. Urban forests include many different environments such as city greenbelts; street and utility rights-of-way; forested watersheds of municipal reservoirs; and residential, commercial, and industrial property. An important distinction between urban and rural forestry is that urban trees are more highly valued than rural trees and often receive expensive individual care and attention. Many professional foresters are trained to handle the special problems of urban trees and to foster the diverse benefits they provide.

Sustainable Forest Management

Sustainable forest management is the management of forests according to the principles of sustainable development. Sustainable forest management has to keep the balance between three main pillars: ecological, economic and socio-cultural. Successfully achieving sustainable forest management will provide integrated benefits to all, ranging from safeguarding local livelihoods to protecting the biodiversity and ecosystems provided by forests, reducing rural poverty and mitigating some of the effects of climate change.

The "Forest Principles" adopted at The United Nations Conference on Environment and Development (UNCED) in Rio de Janeiro in 1992 captured the general international understanding of sustainable forest management at that time. A number of sets of criteria and indicators have since been developed to evaluate the achievement of SFM at the global, regional, country and management unit level. These were all attempts to codify and provide for independent assessment of the degree to which the broader objectives of sustainable forest management are being achieved in practice. In 2007, the United Nations General Assembly adopted the Non-Legally Binding Instrument on all types of forests. The instrument was the first of its kind, and reflected the strong international commitment to promote implementation of sustainable forest management through a new approach that brings all stakeholders together.

A definition of SFM was developed by the Ministerial Conference on the Protection of Forests in Europe (Forest Europe), and has since been adopted by the Food and Agriculture Organization (FAO). It defines sustainable forest management as:

"The stewardship and use of forests and forest lands in a way, and at a rate, that maintains their biodiversity, productivity, regeneration capacity, vitality and their potential to fulfill, now and in the future, relevant ecological, economic and social functions, at local, national, and global levels, and that does not cause damage to other ecosystems".

In simpler terms, the concept can be described as the attainment of balance – balance between society's increasing demands for forest products and benefits, and the preservation of forest health and diversity. This balance is critical to the survival of forests, and to the prosperity of forest-dependent communities.

For forest managers, sustainably managing a particular forest tract means determining, in a tangible way, how to use it today to ensure similar benefits, health and productivity in the future. Forest managers must assess and integrate a wide array of sometimes conflicting factors – commercial and non-commercial values, environmental considerations, community needs, even global impact – to produce sound forest plans. In most cases, forest managers develop their forest plans in consultation with citizens, businesses, organizations and other interested parties in and around the forest tract being managed. The tools and visualization have been recently evolving for better management practices.

The Food and Agriculture Organization of the United Nations, at the request of Member States, developed and launched the Sustainable Forest Management Toolbox in 2014, an online collection of tools, best practices and examples of their application to support countries implementing sustainable forest management.

Because forests and societies are in constant flux, the desired outcome of sustainable forest management is not a fixed one. What constitutes a sustainably managed forest will change over time as values held by the public change.

Criteria and Indicators

Deforestation of native rain forest in Rio de Janeiro City for extraction of clay for civil engineering.

Criteria and indicators are tools which can be used to conceptualise, evaluate and implement sustainable forest management. Criteria define and characterize the essential elements, as well as a set of conditions or processes, by which sustainable forest management may be assessed. Periodically measured indicators reveal the direction of change with respect to each criterion.

Criteria and indicators of sustainable forest management are widely used and many countries produce national reports that assess their progress toward sustainable forest management. There are nine international and regional criteria and indicators initiatives, which collectively involve more than 150 countries. Three of the more advanced initiatives are those of the Working Group on Criteria and Indicators for the Conservation and Sustainable Management of Temperate and Boreal Forests (also called the Montréal Process), Forest Europe, and the International Tropical Timber Organization. Countries who are members of the same initiative usually agree to produce reports at the same time and using the same indicators. Within countries, at the management unit level, efforts have also been directed at developing local level criteria and indicators of sustainable forest management. The Center for International Forestry Research, the International Model Forest Network and researchers at the University of British Columbia have developed a number of tools and techniques to help forest-dependent communities develop their own local level criteria and indicators. Criteria and Indicators also form the basis of third-party forest certification programs such as the Canadian Standards Association's Sustainable Forest Management Standards and the Sustainable Forestry Initiative Standard.

There appears to be growing international consensus on the key elements of sustainable forest management. Seven common thematic areas of sustainable forest management have emerged based on the criteria of the nine ongoing regional and international criteria and indicators initiatives. The seven thematic areas are:

- Extent of forest resources.
- Biological diversity.

- Forest health and vitality.
- Productive functions and forest resources.
- Protective functions of forest resources.
- Socio-economic functions.
- Legal, policy and institutional framework.

This consensus on common thematic areas (or criteria) effectively provides a common, implicit definition of sustainable forest management. The seven thematic areas were acknowledged by the international forest community at the fourth session of the United Nations Forum on Forests and the 16th session of the Committee on Forestry. These thematic areas have since been enshrined in the Non-Legally Binding Instrument on All Types of Forests as a reference framework for sustainable forest management to help achieve the purpose of the instrument.

On January 5, 2012, the Montréal Process, Forest Europe, the International Tropical Timber Organization, and the Food and Agriculture Organization of the United Nations, acknowledging the seven thematic areas, endorsed a joint statement of collaboration to improve global forest related data collection and reporting and avoiding the proliferation of monitoring requirements and associated reporting burdens.

Ecosystem Approach

The Ecosystem Approach has been prominent on the agenda of the Convention on Biological Diversity (CBD) since 1995. The CBD definition of the Ecosystem Approach and a set of principles for its application were developed at an expert meeting in Malawi in 1995, known as the Malawi Principles. The definition, 12 principles and 5 points of "operational guidance" were adopted by the fifth Conference of Parties (COP5) in 2000. The CBD definition is as follows.

The ecosystem approach is a strategy for the integrated management of land, water and living resources that promotes conservation and sustainable use in an equitable way. Application of the ecosystem approach will help to reach a balance of the three objectives of the Convention. An ecosystem approach is based on the application of appropriate scientific methodologies focused on levels of biological organization, which encompasses the essential structures, processes, functions and interactions among organisms and their environment. It recognizes that humans, with their cultural diversity, are an integral component of many ecosystems.

Sustainable forest management was recognized by parties to the Convention on Biological Diversity in 2004 (Decision VII/11 of COP7) to be a concrete means of applying the Ecosystem Approach to forest ecosystems. The two concepts, sustainable forest management and the ecosystem approach, aim at promoting conservation and management practices which are environmentally, socially and economically sustainable, and which generate and maintain benefits for both present and future generations. In Europe, the MCPFE and the Council for the Pan-European Biological and Landscape Diversity Strategy (PEBLDS) jointly recognized sustainable forest management to be consistent with the Ecosystem Approach in 2006.

National Forest Funds

The development of National Forest Funds is one way to address the issue of financing sustainable forest management. National forest funds (NFFs) are dedicated financing mechanisms managed by public institutions designed to support the conservation and sustainable use of forest resources. As of 2014, there are 70 NFFs operating globally.

Forest Genetic Resources

Appropriate use and long-term conservation of forest genetic resources (FGR) is a part of sustainable forest management. In particular when it comes to the adaptation of forests and forest management to climate change. Genetic diversity ensures that forest trees can survive, adapt and evolve under changing environmental conditions. Genetic diversity in forests also contributes to tree vitality and to the resilience towards pests and diseases. Furthermore, FGR has a crucial role in maintaining forest biological diversity at both species and ecosystem levels.

Selecting carefully the forest reproductive material with emphasis on getting a high genetic diversity rather than aiming at producing a uniform stand of trees, is essential for sustainable use of FGR. Considering the provenance is crucial as well. For example in relation to climate change, local material may not have the genetic diversity or phenotypic plasticity to guarantee good performance under changed conditions. A different population from further away, which may have experienced selection under conditions more like those forecast for the site to be reforested, might represent a more suitable seed source.

Forest Ecosystem Management

Forestry must balance the short-term needs and desires of today's human population, the anticipated needs of future generations, and the maintenance of long-term forest ecosystem conditions, functions and organisms. The most effective way to satisfy these obligations may be to use the paradigm of forest ecosystem management (FEM) as the template for forestry.

FEM requires long-term tenure and the management of all desired values and services under an integrated management plan that spans at least one full rotation (multiple rotations for short-rotation timber crops) and has inputs from multiple stakeholders.

An FEM plan should cover a defined forest area sufficiently large to permit the management of key ecological processes. All marketable values should be managed under a single plan to facilitate value trade-off analysis, and targets for the management of non-marketable values should be an integral part of the plan rather than merely a constraint on the profits from a single marketable value such as timber, as is often the case.

These ideas are not new, but have proven difficult to implement. The following are fundamental to achieving FEM: appropriate tenure systems of adequate duration and area, the involvement of multiple stakeholders in setting management objectives, active management and marketing of multiple values at the landscape scale, and the explicit definition of a "desired forest future". The tenure system must permit the management of the entire forest landscape ecosystem as a single endeavour, with economic rewards to the managing agency, consortium or collective for all marketable values. Equally fundamental is the need to understand and respect the ecology of the resources and other values that are to be sustained.

Another major impediment to implementation of FEM has been the lack of ecologically-based scenario and value trade-off analysis tools with which to evaluate the most effective stand- and landscape- management strategies to achieve a desired forest future. Such tools must incorporate both the ecosystem processes responsible for biophysical sustainability, and the social values that relate to social and economic sustainability. In the absence of these prerequisites for successful FEM, this urgent environmental necessity may remain an ecotopian ideal rather than a practical reality.

The world's forests will have to service the needs and desires of another three to four billion people by the end of this century. The current population of about 6.3 billion is posing unprecedented pressure on the world's forests, especially in developing countries at tropical and subtropical latitudes. A plethora of new paradigms have been suggested as templates for managing the world's forests in the face of this human pressure.

Forestry is constantly facing new social pressures, ideas, buzzwords and paradigms. There is an urgent need for clearer definitions of suggested new paradigms, including FEM, and for the ability to forecast their social and environmental consequences over both short and long time scales, and small and large spatial scales, before they are implemented. Only when we are equipped with these forecasts can we have confidence that the suggested new approaches to forest management will honour intergenerational equity issues, and not just indulge short-term desires and preferences.

Forestry is the art (skill), practice, science and business of managing forest stands and forested landscapes to sustain the balance of values and environmental services that are ecologically possible and desired by society. As society has come to understand more about the holistic, integrated character of forest ecosystems, it is demanding ecosystem management as a replacement for traditional timber or other single-value management. It is the responsibility of the profession to make this transition once it is clear what is meant by FEM, but only after it has been assessed that the components of the proposed definition of FEM will respect the ecology of the forests and values in question, and are likely to satisfy the needs of both local people and society in general, now and in the future. Once this has been achieved, foresters will face the difficult challenge of persuading the public to accept what needs to be done to achieve FEM.

The term FEM has seen increasing usage over the past two decades. From the outset, it has been poorly defined, ambiguous and subject to diverse interpretations. The most recent evolution of terminology associated with the FEM approach to forest stand management is Variable Retention silviculture and Structure Management silviculture.

Forest Ecosystem

Forest ecosystems are areas of the landscape that are dominated by trees and consist of biologically integrated communities of plants, animals and microbes, together with the local soils (substrates) and atmospheres (climates) with which they interact. Forests are much more than the present population or community of trees. Forests that have been recently killed or altered by fire, insects, disease, wind or logging are still forests because of the biological and physical legacies from the previous forest - legacies of forest soil, organic matter, microbes, minor vegetation and animals. Under a regime of sustainable forest management, many or most of these legacies persist during the period between forest disturbance and the redevelopment of tree cover.

Forest ecosystems are both a stand-level and a landscape phenomenon, the latter being a mosaic of stands that vary in age, species composition, structure, function and time since disturbance. Periodic disturbance is a key attribute of most forest ecosystems, and maintenance of their historical character and values will generally require maintenance of historical disturbance regimes, or the ecological effects thereof.

Because a forest ecosystem is an integrated biophysical system, a forest is as much a set of ecosystem processes as a set of forest ecosystem components. Short-term changes in the structure of the forest do not constitute loss of the forest, as long as the processes of the forest ecosystem remain in operation at acceptable levels.

FEM approaches forest conservation, utilization, administration and regulation on the basis that the forest is a highly integrated, complex, generally resilient, multi-value biophysical system that has thresholds of tolerance for disturbance (either too much or too little) beyond which its resilience and certain values and environmental services are changed, and often reduced. FEM is the management of forest ecosystem processes and disturbance regimes to sustain the desired values and ecosystem services from a shifting mosaic of different ecosystem conditions across the landscape, and a non-declining pattern of change over time in the values and services provided by each stand in that landscape. It is also the management of the human use of, and interactions with the forest, because humans are part of forest ecosystems.

Based on these definitions, it is clear that FEM is far more than a politically correct buzzword. However, if we are to be successful in implementing it as the biophysical template for sustainable, multi-value forestry, we must understand the impediments to, and the necessary conditions for its successful application. Successful application of the FEM concept as the basis for forestry involves much more than mere acceptance that FEM is a nice idea.

Elements of Forest Ecosystem Management

It includes the following elements:

- Sustainability: FEM sees intergenerational sustainability as a precondition.
- Goals: FEM sets measurable goals that specify future processes and outcomes necessary for sustainability.
- Sound ecological models and understanding: FEM relies on research performed at all levels of ecological organization.
- Complexity and connectedness: FEM recognizes that biological diversity and structural complexity strengthen ecosystems against disturbance and supply the genetic resources necessary to adapt to long-term change.
- The dynamic character of ecosystems: FEM avoids attempts to "freeze" ecosystems in a particular state or configuration.
- Context and scale: FEM recognizes that ecosystem processes operate over a wide range of spatial and temporal scales, and their behaviour at any given location is greatly affected by surrounding systems.
- Humans as ecosystem components: FEM values the active role of humans in achieving sustainable management goals.
- Adaptability and accountability: FEM acknowledges that current knowledge and paradigms of ecosystem function are provisional, incomplete and subject to change.

There are several different lists of the key themes in forest ecosystem management. The following is a distillation of those lists:

- Establish a clear set of management goals that respect the ecological tolerances and resilience of the ecosystems, species and values of interest.
- Manage forests as ecosystems, crossing levels of biological organization and integration.
- Define management areas ecologically, and vary the practices of management to respect the ecological and biological diversity therein.

- Assign the management of forest ecosystems at both the stand and landscape scales, over a defined forest area and for ecologically appropriate time scales, to a single agency or organization under a single plan that has the mandate to manage for a sustained balance of values across the landscape and the temporal variation in these values in any one stand.
- Maintain ecosystem function, seral sequences, and species composition within historical ranges of variation (emulation of natural forest disturbance) or a socially acceptable subset thereof.
- Manage adaptively, constantly learning and changing practices as indicated by monitored outcomes, and using ecologically based forecasting tools to support management planning where experience does not yet provide the needed guidance.

Prerequisites for Forest Ecosystem Management

There are many conditions that must be met before FEM can become an effective paradigm for forestry. These include:

- Appropriate tenure systems and management plans: Forests can only be managed as ecosystems if all values are managed under a single, integrated, multi-value and ecologically based plan by a single agency/organization/collective over time scales that are consistent with the ecosystem processes responsible for resilience and sustainability. The tenure and plan should also cover an area large enough to encompass the key ecosystem processes. The objectives of the plan must be established by a multi-stakeholder process.
- Establishment of overall long-term objectives: There must be an explicit definition of a "desired forest future". Only when the management plan has a statement of the full range of values and conditions that are to be available at various times in the future can there be a plan as to how to achieve this. Short-term values are important, but these must be balanced against the achievement of long-term goals.
- Adequate inventory: An FEM plan must be based on an inventory of ecological, biological and social diversity.
- Acceptance of the dynamic nature of ecosystems: A key prerequisite for FEM is acceptance of the dynamic nature of forest ecosystems, and the role of ecosystem disturbance in the long-term maintenance of desired values.
- Founding FEM on the concept of "ecological rotations": Because of the vital ecological role of ecosystem disturbance, and because forest management involves management of "natural" and human-caused disturbance, sustainable FEM should be founded on the concept of "ecological rotations". An ecological rotation is that combination of severity, scale, pattern and frequency of ecosystem disturbance and ecosystem resilience (speed of return to pre-disturbance

condition, or to some desired new condition) that results in non-declining patterns of stand-level change, and a shifting landscape mosaic of acceptably constant overall character. The ecological rotation concept provides foresters with choices. If rotations are to be short for economic or other reasons, the higher frequency of disturbance may need to be matched by lower severity and scale of disturbance. Where higher levels of disturbance are required for ecological or social reasons, the frequency of disturbance should be reduced (longer rotations). Alternatively, management interventions can be made to increase the resilience of the ecosystem (e.g. planting, competition control, fertilization) so that more frequent and more severe disturbance remains sustainable. Different values have different ecological rotations, which suggests that variable retention silviculture (where different elements within the ecosystem are managed on different ecological rotations) should be a major component of ecological rotation-based forestry. Selection of that balance of frequency, severity, scale and management of resilience that meets the objectives of FEM most effectively is a challenging task. It calls for ecologically based planning tools - ecosystem-management models - capable of dealing with ecosystem processes, multiple values, the spatial and temporal scales involved, and the interactive effects of severity, scale, frequency of disturbance and ecosystem resilience.

- Economic sustainability: Unless forestry is economically viable it may revert to exploitation, which is not forestry. FEM is based on the concept that humans are part of and dependent on forest ecosystems, and therefore in self-interest must manage them sustainably, but this is expensive. Exploitation is cheaper in the short term. Therefore, the endeavour must be economically viable. Where the economics of forestry are marginal, over-cutting may occur. On the other hand, where the economics of stand thinning prevents this management practice, stands may become more "natural", increasing values that are dependent on the accumulation of dead trees in the ecosystem. By marketing non-timber as well as timber forest products - a major feature of FEM - forestry that is uneconomic based on timber alone may become economic.

- Communication and planning tools: Establishment of ecologically and socially achievable objectives, and evaluation of the most acceptable way of satisfying these, requires planning tools that have capabilities far in excess of those traditionally used in forestry, and an improved capability to communicate the output of such tools to the public.

Ecosystem Management Models as Communication and Planning Decision-support Tools

Setting objectives, assessing trade-offs between values, balancing the needs of the present generation against the legacies we want to pass to future generations, and communicating choices, risks, uncertainties and possible outcomes to the public all require the use of a new generation of decision-support tools. Traditional planning tools will continue to be essential. These range from forest site classifications, to geographical information systems, to timber supply-analysis tools. However, the increasing complexity of forestry, and the imperative to manage forest ecosystems rather than populations of trees or individual games species, requires decision-support tools that can address this complexity. This can be provided by FEM models. These should be linked to user-friendly interfaces and visualization systems to facilitate their use by the non-technical population.

Forest decision-support tools fall into three major categories: experience-based, empirical (historical bioassay [HB] models; e.g. growth and yield models); knowledge-based (process simulation [PS] models); and hybrid simulation [HS] models that combine both experience and knowledge in hybrid historical bioassay-process simulation systems. The emerging consensus is that HB will always be useful in forestry but have limited capabilities relative to changing expectations of such tools, that PS models are needed but that they are often too complex to use in forest management, and that a hybrid of the two - HS models - will be the way of the future.

The public are generally better able to understand and evaluate forestry through pictorial than through tabular, graphical and two-dimensional map representations. The scientific basis for the images being presented must be communicated in as clear a format as possible in order to establish confidence in the images. Visualizations should incorporate representations of natural and human-caused risks and uncertainties, and clearly communicate that there are many different possible forest futures.

An important question with respect to forestry models is: can they be used by foresters in developing countries? While it is true that there are knowledge and technology barriers that limit the use of such tools in many parts of the world at present, computers are rapidly becoming omnipresent wherever there is electricity. Computer models can harness knowledge and experience from developed countries at any latitude, and make them available as heuristic, educational and extension tools to countries that presently lack the wherewithal to locally calibrate and use such tools. It may be some years or decades before foresters in the less developed countries gain the local information needed to make them realistic local tools for prediction, but the day when this will be possible is approaching, and it is time to start to customize this type of tool for this future use. Several forest ecosystem models already exist for tropical forests, and there are many for temperate forests. Their development should be continued in anticipation of the day that foresters in developing tropical and temperate countries are ready for them.

Ecoforestry

Ecoforestry has been defined as selection forestry or restoration forestry. The main idea of Ecoforestry is to maintain or restore the forest to standards where the forest may still be harvested for products on a sustainable basis. Ecoforestry is forestry that emphasizes holistic practices which strive to protect and restore ecosystems rather than maximize economic productivity. Sustainability of the forest also comes with uncertainties. There are other factors that may affect the forest furthermore than that of the harvesting. There are internal conditions such as effects of soil compaction, tree damage, disease, fire, and blow down that also directly affect the ecosystem. These factors have to be taken into account when determining the sustainability of a forest. If these factors are added to the harvesting and production that comes out of the forest, then the forest will become less likely to survive, and will then become less sustainable.

Since the forest is considered an ecosystem, it is dependent on all of the living and non-living factors within itself. This is a major part of why the forest needs to be sustainable before it is harvested. For example, a tree, by way of photosynthesis, converts sunlight to sugars for respiration to keep the tree alive. The remains of the converted sugars is left in roots for consumption by the organisms surrounding the tree in the habitat. This shows the productivity of an ecosystem with its inhabitants. Productivity within the ecosystem cannot come to fruition unless the forest is sustainable enough to be harvested. If most individual organisms of the ecosystem vanish, the ecosystem itself is at risk. Once that happens, there is no longer any forest to harvest from. The overall productivity of a system can be found in an equation where the Net Primary Production, or NPP, is equal to the Gross Primary Production, or GPP, minus the Respiration, or R. The formula is the NPP = GPP - R. The NPP is the overall efficiency of the plants in the ecosystem. Through having a constant efficiency in NPP, the ecosystem is then more sustainable. The GPP refers to the rate of energy stored by photosynthesis in plants. The R refers to the maintenance and reproduction of plants from the energy expended.

Ecoforestry has many principles within the existence of itself. It covers sustainable development and the fair harvesting of the organisms living within the forest ecosystem. There have been many proposals of principles outlined for ecoforestry. They are covered over books, articles, and environmental agencies. All of the principles relate to the idea that in ecoforestry, less should be harvested, and diversity must be managed. Through harvesting less, there is enough biomass left in the forest, so that the forest may stay healthy and still stay maintained. It will grow at a sustainable level annually, and thus it will be able to still be harvested the following year. Through management of the diversity, species may cohabitate in an ecosystem where the forest may feed off of other species in its growth and production. The Principles of Ecoforestry may be found below.

Principles of Ecoforestry

The principles of ecoforestry are:

- One must consider retention: It must be the first consideration in any planned removal of trees from a stand. By placing retention at the beginning of a planned removal, one makes sure that they know what must be left to ensure the protection of such things as rare species, sites of native cultural significance, riparian zones.

- Riparian zones should not be touched: It is a sensitive area, thus tree removal should not occur there. This protects the water quality. Water quality is protected by not altering the drainage pattern of the zones. No tree removal should take place in the most sensitive areas.
- The composition and structures should be upheld so that forests may fully function. This may be subject to large old trees, snags, and large fallen trees. These pieces of the ecosystem are upheld through letting them grow and die out into a timber extraction area.
- When removing trees, the lowest impact should be used: This means not compacting the soil in the forest and building small or no roads if at all possible. This takes away the disturbance that would be done to the foundation of the forest's grounds.
- Plan in terms of the needs of the larger watershed: The watershed zone plan designates areas where tree removal is not permitted. It also designates the areas where removal is possible, and the different types of removal available.
- Prohibit clear cutting: Clear cutting is not ecologically correct. There are methods for cutting trees to maintain structure, and to allow the trees to cohabitate in a healthy ecosystem.
- Select trees as candidates for removal by considering how abundant and redundant their structures and functions are to the rest of the forest as a whole, leaving potential wildlife trees (to become snags and coarse woody debris).
- The forest should be allowed to regenerate trees through the seeds from the trees in the logged areas. This allows tree planting to be taken out of the ecoforestry principles, allowing a natural regeneration.
- Ecological succession should be kept at all times: This will protect biological diversity. This even means eliminating the process of brush control.
- No slash burning: Fire may be used as a tool in landscapes that have a history of naturally occurring fires. Although this may be true, fire should still be used with caution.
- No pesticide use: The forest needs disease, insects, and shrub/herb vegetation. They are essential parts of a fully functioning forest, even though they involve the forest's decay in some cases. They are natural parts that allow the ecosystem to operate as a whole.
- Maintain and restore topsoil quality: This can be done through leaving sufficiently coarse and small debris on the forest grounds.

- Maintain beauty and other natural aesthetic qualities in the visual, sound, and odor landscapes. This includes not taking away the wildlife, the plants, or structural layout of the ecosystem.
- Always look at the forest as a whole: Each part of the forest contributes to its overall needs and health. That is how the forest would have survived without human interference.
- Rely more on people and markets: Use accounting and budgeting as a solution to rely less on the destruction or harvesting from the forests.
- Don't do wrong: If it feels wrong, then it is probably wrong. Don't allow ignorance to persevere. Recognize that the ecosystem needs tending to. If the forest is not preserved, then it cannot be harvested forever.

Forest Governance

In general terms, governance refers to the formal and informal rules, organizations and processes through which public and private actors articulate their interests and make and implement decisions. Forest governance is defined as the way in which public and private actors, including formal and informal institutions, smallholder and indigenous organizations, small, medium-sized and large enterprises, civil-society organizations and other stakeholders negotiate, make and enforce binding decisions about the management, use and conservation of forest resources. The concept of forest governance has evolved to engage multiple (public and private) actors at multiple scales, from local to global. It may include:

- Rules about how forests should be governed, governmental regulations about who benefits from forest resources, and traditional and customary rights.
- The use of private-sector mechanisms such as voluntary certification to support SFM and legal timber supply.
- International measures to support timber legality and promote good governance, such as the European Union's Forest Law Enforcement, Governance and Trade (FLEGT) Action Plan and payment schemes for environmental services.

Effective forest governance processes engage forest stakeholders, address key forest-related issues, and involve other sectors that affect, or are affected by, forest governance.

Among them, women in particular must be taken into account, as they are one of the main groups using forests. It is essential that women be involved in these processes because rural women's dependence on forests is different from, and often greater than, men's due to the gender division of labour and different access to economic resources.

The debate around forest governance has increasingly considered gender, but there is little documented evidence of an actual increase in women participating in the process. In high-level national and international discussions and positions, women are still underrepresented. In community-based approaches, especially in South Asia, the bodies governing and setting the rules for accessing and managing forests, such as a general assembly and executive committee, were so-called "gender neutral". But women's participation was restricted because of conservative gender norms. Even when women participate, they often have only a passive or consultative role, which allows them only to exchange opinions that have no guarantee of influencing decisions.

Analog Forestry

Analog forestry is an approach to ecosystem restoration that considers the process of forest formation and the functioning of forest services to be critical in establishing a sustainable ecosystem characterised by a high biodiversity to biomass ratio. Design is produced through a synthesis of traditional and scientific knowledge. It seeks to optimise the productive potential of the design rather than maximise the production of one crop and to maximise ecosystem services by increasing the volumetric mass of the photosynthetic component.

Analogue forestry draws design input not only from traditional models but also from the natural forest successional dynamics. When an ecosystem is designed to be analogous to the indigenous climax state, the efficiency and dynamics of the natural processes can be replicated. These quasi-natural forests are designed to mimic the structural and functional aspects of indigenous forests and are referred to as analog forests. In addition to their ecological characteristics, analog forests are also designed to provide economic benefits. However, it is not until all the ecological requirements of the location are satisfied that economic values of species are considered. Therefore, an analog forest may comprise natural and exotic species in any proportion, the contribution to structure and function being the overriding factor that determines its use.

The theoretical underpinnings began in 1978 in San Diego and Guatemala. It was first implemented in Sri Lanka around 1981 by Ranil Senanayake as an alternative to monocultures of *Pinus* and *Eucalyptus* and has spread to India, Vietnam, Philippines, Australia, Peru, Ecuador, Colombia, Brazil, Costa Rica, Dominican Republic, Honduras, Mexico, Canada, Kenya and Zimbabwe at present. The International Analog Forestry Network (IAFN) is currently hosted in Costa Rica.

Analog forestry is a system of silviculture that seeks to design, plan and manage ecosystems dominated by woody perennials. It has been primarily employed in tropical or subtropical areas, but can be used in temperate areas too. The design seeks to mimic the architectural structure and ecological function of the preexisting climax vegetation of the area, and can be designed to provide economic, social and environmental benefits.

Analog forestry always considers the inclusion of non-crop native species to be considered in the design and thus addresses the inclusion of native biodiversity in the design. As analog forestry also requires the inclusion of long-lived species of trees in the design it has the capacity to sequester carbon for a longer time than plantation forestry. Analogue forestry has the potential to produce very high values of photosynthetic biomass as its design calls for the inclusion of all the growth forms that occupy the three-dimensional space of the mature indigenous forest. By including many species of crops in one area analog forestry helps spread the risk of market failure on a single crop.

Community Forestry

Community forestry is an evolving branch of forestry whereby the local community plays a significant role in forest management and land use decision making by themselves in the facilitating support of government as well as change agents. It involves the participation and collaboration of various stakeholders including community, government and non-governmental organisations (NGOs). The level of involvement of each of these groups is dependent on the specific community forest project, the management system in use and the region. It gained prominence in the mid-1970s and examples of community forestry can now be seen in many countries including Nepal, Indonesia, Korea, Brazil, India and North America.

National Forest Inventory of Peru.

Community forestry is a branch of forestry that deals with the communal management of forests for generating income from timber and non-timber forest products as forms of goods while in other hand regulating ecosystem, downstream settlements benefits from watershed conservation, carbon sequestration and aesthetic values as in forms of services. It has been considered one of the most promising options of combining forest conservation with rural development and community empowerment and poverty reduction objectives. Community forestry is defined by the Food and Agricultural Organization of the United Nations as "any situation that intimately involves local people in forestry activity". Community forestry exists when the local community in an area plays a significant role in land use decision-making and when the community is satisfied with its involvement and benefits from the management of the surrounding forest and its resources. Community forestry is deepened by SPERI via the concept of spirit forest where is source and space for maintaining livelihood identity and cultural values of local communities.

Community forestry is first implemented through the establishment of a legal and institutional framework including the revision of legal norms and regulations for forest management, the development of National Forest Plans and the strengthening of decentralization processes to sub-national levels of government. The second principal line of action is the implementation of pilot projects to demonstrate the feasibility of the community forestry framework. However, a study by the Overseas Development Institute shows that the technical, managerial and financial requirements stipulated by the framework are often incompatible with local realities and interests. A successful legal and institutional framework will incorporate the strengthening of existing institutions and enable the dissemination of locally appropriate practices as well as the local capacity for regulation and control.

In a 2016 review of community-based forestry, FAO estimated that almost one-third of the world's forest area is under some form of community-based management. Community forestry first came to prominence in the mid-1970s and has continued to evolve over the last few decades in a growing number of countries. The availability of forest resources are often greatly reduced for use by the local people due to increasing pressures to cultivate the land, reliance on the forest resources are also affected by economic and political changes. The evolution of community forestry in Nepal dates back to the late 1970s and was first instilled as an attempt to improve the management of forest resources and address environmental issues that were of great concern with the countries failing centralized forest policy. Over the past two decades, community forestry has been applied successfully in many developing countries, with its main goal being the alleviation of poverty amongst local forest communities and forest conservation. More recently, community forestry has been implemented in developing countries and it has been successful in its aims of sustainable forest management, climate change adaptation plan of action, and securing socio-economic benefits for local communities.

Stakeholders

There is a large variety of stakeholders involved when considering community forestry. Participation from some of the various levels of community, government and non-government organisations (NGOs) are essential in the project's success. While specific stakeholders vary between different community forestry projects the primary stakeholder groups are as follows:

- Local Community:
    - Communities living adjacent to or within the forest.
    - Traditional leaderships including village chief, elderly, clan heads, and spiritual leaderships.
    - Community-based organisations including forest users groups, networks of inter-villages forest protection, etc.
    - Community representatives/local councils.
- Government:
    - State Government.
    - National Government.
    - Departments of Agriculture, Forestry, Environment etc. (specific the country and region).
- NGO's:
    - Environmental/conservation groups.
    - Commercial forestry industries.
    - Industries reliant on forests. (i.e. harvestable products other than wood, e.g. game meat).
    - Tourism industry.
    - Animal welfare groups.

Natural Ecosystem

Stakeholders of community forestry have a vested interest to establish sustainable practices, whether this is to develop and maintain a regular income, ensure that forests are sufficiently protected to ensure their longevity or to reduce illegal activities and manage the area in such a way to promote tourism and conservation. In this situation, stakeholders came to conclusion to handover forest resources to local communities for conserving, managing and utilization by their own decision. Despite significant development, continued improvement in the collaboration between local governments and forest communities seems to be a key point for better community forest management. A wide range of futures scenarios have been put up to help the environmental decision process.

Challenges

A study conducted in the Brazilian Amazon determined that there are a number of challenges that must be faced when developing a sustainable management strategy for community forestry. These challenges show the impact each management challenge has on other obstacles. The model is segregated into two phases: the development phase during which several enabling factors (land ownership, organisational capacity, technical knowledge and capital) are needed to obtain a legal management permit and secondly the operational phase where factors (clandestine loggers, access to markets, infrastructure and managerial skills) influence the successfulness of the management program. Each of the challenges must be addressed in order for a self-sustaining community forestry management program to be established.

Challenges faced by stakeholders managing forests:

- Land ownership: It is not uncommon for small settlements living traditionally to lack clear title to the forests and lands; as such their rights to harvest the land may come into dispute. Formal recognition of land ownership or rights to use is needed for legal forest management. In Nepal, there is still confusion to communities about their tenure into services such as income from trading of carbon sequestration come from forestry resources. Government wants to hold revenue raised from trading of carbon but communities claim that it is our property due to be managed by us.
- Organizational capacity: Organizational capacity refers to the community's competence to organize a forest management program.
- Capital: A community will require start-up capital to invest in the required infrastructure, equipment, and to hire a forester to organise and oversee management plans.
- Technical knowledge: While it is common for community members have a thorough understanding of forest ecology in a natural and historical sense, they often lack the technical knowledge and legal certification required to manage forest resources legally.
- Regulatory obstacles: The forest sector generally has been over-regulated, not under-regulated. In contrast other sectors are generally less regulated and more profitable (e.g. agriculture, mining, tourism, real estate), and conversion is often a less complex and more financially rewarding option than sustainable forest management. An example of this is the forest management plans (FMPs), which tend to be the principal instruments for regulatory compliance. Typically, FMPs have been (and remain) complex, costly, and difficult to develop, implement, monitor, and enforce. In addition, as they rarely consider traditional or local knowledge in promoting sustainability, they make it even more difficult for forest communities to appropriate them or participate in their execution.

- Legal management: Community forestry needs to be based on a legal management plan, prepared and approved by the relevant government authority (usually state environment agencies). Approval of the management plan can often be a long, bureaucratic process.
- Clandestine loggers: Clandestine (illegal) loggers can enter a region and illegally log valuable species without the community's consent, and without a legal management plan. Illegal harvesting often significantly degrades the forest as few limitations are observed by the illegal loggers. Communities are either robbed of their valuable timber, or are paid below market prices.
- Market access: Small villages that are often the focus of community forestry initiatives generally have limited access to markets due to their physical isolation, precarious transport and communication, limited contact with buyers and lack of marketing knowledge. Often they will also face difficulties competing with large-scale operations and illegal timber that flood the market with cheaper products.
- Infrastructure: Forest management, for all types of enterprise, requires a certain amount of physical infrastructure, such as roads, logging equipment, buildings for storage and management headquarters and a reliable power supply.
- Managerial skills: Skills related to effective management of the program, business knowhow and entrepreneurial ability, day-to-day decision-making, marketing skills, ability to resolve internal conflicts, and ensuring community benefit sharing are often lacked in rural villages.
- Economic returns: The result of inadequacies in management is small and often insufficient economic returns to keep the program afloat, and furthermore to keep the community interested in the activity.

Best Practices

For a conservation program within a Community forest to succeed, community involvement is imperative. Governments with interest in forest conservation introduce state-wide policies and legislations which have historically failed to deliver the desired outcomes such as in China, Nepal and Peru. Moreover, no single stakeholder by itself can ensure the success of such a program.

Continuous Cover Forestry

Continuous cover forestry is an approach to the sustainable management of forests whereby forest stands are maintained in a permanently irregular structure, which is created and sustained through the selection and harvesting of individual trees. The term "continuous cover forestry" does not equate exactly to any one particular silvicultural system, but is typified by selection systems. For example coppice with standards and Reiniger's Target diameter harvesting are also continuous cover forestry. Different existing forest stands may require different silvicultural interventions to achieve a continuously productive irregular structure.

85 year old stand of Douglas fir in the process of transformation to a continuous cover forest.

The term continuous cover forestry has been widely adopted in British forestry practice following the creation of the Continuous Cover Forestry Group in 1991. Currently the UK Forestry Standard and the guidelines of the silvicultural-systems UK Woodland Assurance Standard call for wider use and application of CCF in British forestry, as there is evidence that continuously productive irregular structured woodlands are more robust and resilient in response to climate change, and forest health threats in addition to a wide range of other social and environmental benefits. However, there remains some lack of understanding of the definition and practice of CCF among forestry professionals in Britain, and Ireland.

CCF is an approach to forest management which respects the characteristics and processes inherent to the site, and will normally involve a mixture of tree species and ages. In French, it is referred to as sylviculture irrégulière, continue et proche de la nature (SICPN) (i.e., continuous, irregular and close to nature silviculture), and in German as Dauerwald, or naturgemässe waldwirtschaft (close to nature forestry). Management is based on the selection and promotion of individual trees, of all sizes, rather than the creation of areas with trees of uniform size. Any monitoring of the development of irregular stand structures will be achieved through periodic re-measurement of stem diameters in sample plots or across whole stands. Repeated interventions, typically at intervals of between 3 and 7 years, focus on the removal of trees which are interfering with the growth of trees with greater potential, the harvesting of crop trees at their optimum size, and the promotion of regeneration and small trees, such that an irregular stand structure is maintained in perpetuity.

Regeneration will normally be by the growth of naturally-regenerated seedlings rather than by planting. Artificial regeneration (especially enrichment planting) may be needed to diversify the genetic and species composition, and to increase ecological resilicncc to threats from climate change, pests, or disease.

Principles

The general approach to CCF can be outlined in terms of the following underlying principles, as defined by the Continuous Cover Forestry Group (CCFG):

- Adapt the forest to the site: The forest manager works with the site and respects inherent variation, rather than imposing artificial uniformity.
- Adopt a holistic approach: The whole ecosystem is regarded as the production capital of the forest, including soil, carbon, water, fungi, flora, fauna, as well as the trees themselves.
- Maintain forest habitats: Maintenance of the forest habitat is essential (which requires that clear-felling is avoided).
- Develop the forest structure: Stand improvement is concentrated upon the development of preferred individual trees rather than the creation of a uniform block of stems. A characteristic of permanently irregular stands is that yield control is based upon measurement of stem diameter and increment rather than age and area.

References

- Forest-management-planning, our-natural-resources-forests-forestry-sustainable-forest-management-17493: nrcan.gc.ca, Retrieved 02 January, 2020
- Matta, Rao (2015). Towards effective national forest funds, FAO Forestry Paper 174 (PDF). Rome, Italy: Food and Agriculture Organization of the United Nations. ISBN 978-92-5-108706-0
- Don, Gilmour (2016). FAO Forestry Paper 176: Forty years of community-based forestry: A review of its extent and effectiveness (PDF). FAO. ISBN 978-92-5-109095-4. Retrieved 5 April, 2016
- Purposes-and-techniques-of-forest-management, science-forestry: britannica.com, Retrieved 08 August, 2020
- Senanayake, R.; John, J, Jack (1998). Analog Forestry: An Introduction. Monash University Publications. Monash Univ.Clayton, Vic. Australia. ISBN 978-0909685584

4
Agroforestry

Agroforestry is the combination of agriculture and forestry. It is a land-use system in which shrubs and trees are grown around pastureland to increase biodiversity and reduce erosion. Forest farming and gardening, mycoforestry, etc. fall within it. All the aspects related to agroforestry have been carefully analyzed in this chapter.

Agroforestry is the management and integration of trees, crops and livestock on the same plot of land and can be an integral component of productive agriculture. It may include existing native forests and forests established by landholders. It is a flexible concept, involving both small and large-sized land holdings.

Scientifically speaking, agroforestry is derived from ecology and is one of the three principal land-use sciences, the other two being agriculture and forestry. Agroforestry differs from the latter two principals by placing an emphasis on integration of and interactions among a combination of elements rather than just focussing on each element individually.

Agroforestry has a lot in common with intercropping (the practice of planting two or more crops on the same plot) with both practices placing an emphasis on interaction between different plant species. Generally speaking, both agroforestry and intercropping can result in higher overall yields and reduced operational costs.

Benefits of Agroforestry

Over the past two decades, a number of studies have been carried out analysing the viability of agroforestry. The combined research has highlighted that agroforestry can reap substantial benefits both economically and environmentally, producing more output and proving to be more sustainable than forestry or agricultural monocultures. Agroforestry systems have already been adopted in many parts of the world.

Agroforestry systems can include the following benefits:

- They can control runoff and soil erosion, thereby reducing losses of water, soil material, organic matter and nutrients.
- They can maintain soil organic matter and biological activity at levels satisfactory for soil fertility. This depends on an adequate proportion of trees in the system- normally at least 20% crown cover of trees to maintain organic matter over systems as a whole.
- They can maintain more favourable soil physical properties than agriculture, through organic matter maintenance and the effects of tree roots.
- They can lead to more closed nutrient cycling than agriculture and hence to more efficient use of nutrients. This is true to an impressive degree for forest garden/farming systems.
- They can check the development of soil toxicities, or reduce exiting toxicities-both soil acidification and salinization can be checked and trees can be employed in the reclamation of polluted soils.
- They utilize solar energy more efficiently than monocultural systems different height plants, leaf shapes and alignments all contribute.
- They can lead to reduced insect pests and associated diseases.
- They can be employed to reclaim eroded and degraded land.

- Agro forestry can augment soil water availability to land use systems. In dry regions, though, competition between trees and crops is a major problem.
- Nitrogen-fixing trees and shrubs can substantially increase nitrogen inputs to agro forestry systems.
- Trees can probably increase nutrient inputs to agro forestry systems by retrieval from lower soil horizons and weathering rock.
- The decomposition of tree and pruning can substantially contribute to maintenance of soil fertility. The addition of high-quality tree prunings leads to large increase in crop yields.
- The release of nutrients from the decomposition of tree residues can be synchronized with the requirements for nutrient uptake of associated crops. While different trees and crops will all have different requirement, and there will always be some imbalance, the addition of high quality prunings to the soil at the time of crop planting usually leads to a good degree of synchrony between nutrient release and demand.
- In the maintenance of soil fertility under agro forestry, the role of roots is at least as important as that of above-ground biomass.
- Agro forestry can provide a more diverse farm economy and stimulate the whole rural economy, leading to more stable farms and communities. Economics risks are reduced when systems produce multiple products.

As well as building on practices used in forestry and agriculture, agroforestry also works towards land protection and conservation through more effective protection of stock, control of soil erosion, salinity and water tables and a higher quality control of timber.

A denser, more-dependable tree covering can provide shelter to livestock during the warmer months allowing the animals can conserve energy. That same tree covering helps block out wind, helping to boost water retention levels that can help produce a more robust crop yield.

Agroforestry can improve land protection in the following areas:

Salinity and Water Table Control

Salinity is mainly caused by rising water tables. Trees help to lower water tables, acting as pumps to take up water from the soil and then evaporating it to the atmosphere.

Soil Erosion Control

Soil erosion or loss results from the action of wind and water on unprotected soils. The forest canopy, roots and leaf litter all have a role in controlling soil erosion.

Water Logging

Through water removal, established trees can substantially reduce water logging in their immediate area, which may result in improved land uses, e.g. pasture or crop.

Agroforestry can have immense benefits for the environment and the For farmers, the ability to maintain some sort of control over land and production in the face of climate change means agroforestry could hold huge promise for the agricultural sector.

On an environmental level, agroforestry's ability to help prevent soil erosion while simultaneously aiding water retention and promoting soil fertility could help provide a solution for areas where rainfall is irregular or might become irregular due to climate change while dense plantations of trees would also help absorb CO_2 and regulate local temperature.

Forest Farming

Forest farming is the cultivation of high-value crops under the protection of a managed tree canopy. In some parts of the world, this is called multi-story cropping and when used on a small scale in the tropics it is sometimes called home gardening. It is not just recreational harvesting or wild harvesting wild harvesting of native understory wood land plants without management; management is an essential part of forest farming. This approach to crop production intentionally uses both vertical space and the interactions of the plants and microclimate.

The intensity of forest farming production can vary depending on the producer's goals, available markets, processing equipment and the site. Some examples of management activities include harvesting and scattering local seed; thinning out competing plants; additional site preparation for planting seeds, bulbs or plant starts; soil amendments for pH or fertility; constructed raised planting beds; pest control; and even fencing to keep out animals and poachers.

Non-timber specialty products like goldenseal can be cultivated beneath a tree canopy.

Often these woodland crops grown under a canopy are called Non-Timber Forest Products or NTFPs. Crops like ginseng, goldenseal, shiitake or other mushrooms, and decorative ferns are used or sold for medicinal, culinary, and ornamental uses. Forest farming can provide shorter-term income while high-quality trees are being grown for wood or other tree products.

Forest farming is most often used on private lands to supplement family income. Because of the demand and high value of some of these plants they can no longer be found in many places within their historic native range. Wild harvesting of these plants is not allowed on public land in many areas, making forest farming an important option for meeting demand for these plants. Even if the plants of interest are not present in woodland, the careful inspection of the understory woodland plants may identify other non-economic plants that are often found with these more desirable NTFPs. If so, this may indicate a good location to reintroduce and grow the desired NTFP. Additional benefits of this intensified production is the regular attention in the woods may help to spot and control invasive plants and pests as well as the reduced harvest pressure on wild plant populations that allows them to be re-established in other woodlands in the region.

Principles

Forest farming principles constitute an ecological approach to forest management. Forest resources are judiciously used while biodiversity and wildlife habitat are conserved. Forest farms have the potential to restore ecological balance to fragmented second growth forests through intentional manipulation to create the desired forest ecosystem.

In some instances, the intentional introduction of species for botanicals, medicinals, food or decorative products is accomplished using existing forests. The tree cover, soil type, water supply, land form and other site characteristics determine what species will thrive. Developing an understanding of species/site relationships as well as understanding the site limitations is necessary to utilize these resources for production needs, while conserving adequate resources for the long-term health of the forest.

Apart from the environmental benefits, forest farming can increase the economic value of forest property and provide short- and long-term benefits to the landowner. Forest farming provides economic return from intact forest ecosystems, but timber sales can remain part of the long-term management strategy.

Methods

Forest farming methods may include: Intensive, yet careful thinning of overstocked, suppressed tree stands; multiple integrated entries to accomplish thinning so that systemic shock is minimized; and interactive management to maintain a cross-section of healthy trees and shrubs of all ages and species. Physical disturbance to the surrounding area should be minimized.

Level of Management Required

- Forest gardening is the most intensive of forest farming methods. In addition to thinning the overstory, this method involves clearing the understory of undesirable vegetation and other practices that are closely related to agronomy (tillage, fertilization, weeding, and control of disease and insects and wildlife management). Due to input levels, this method often produces lower valued products compared to other methods. Forest gardens take advantage of the vertical levels of light availability and space under the forest canopy so that more than one crop can be grown at once if desired.

- Wild-simulated seeks to maintain a natural growing environment, yet enriches local NTFP populations to create an abundant renewable supply of the products. Minimal disturbance and natural growing conditions ensure products will be similar in appearance and quality of those harvested from the wild. Rather than till, practitioners often rake leaves to expose soil, sow seed directly onto the ground, and then cover with leaves again. Since this method produces NTFPs that closely resemble wild plants; they often command a higher price than NTFPs produced using the forest gardening method.

- Forest tending involves adjusting tree crown density to manipulate light levels that favor natural reproduction of desirable NTFPs. This low intensity management approach does not involve supplemental planting to increase populations of desired NTFPs.

- Wildcrafting is the harvesting of naturally growing NTFPs. It is not considered a forest farming practice since there is no human involvement in the plant's establishment and maintenance. However, wild crafters often take steps to protect NTFPs with future harvests in mind. It becomes agroforestry once forest thinnings, or other inputs, are applied to sustain or maintain plant populations that might otherwise succumb to successional changes in the forest. The most important difference between forest farming and wildcrafting is that forest farming intentionally produces NTFPS, whereas wildcrafting seeks and gathers from naturally growing NTFPs.

Production Considerations

Forest farming can be a small business opportunity for landowners and requires careful planning, including a business and marketing plan. Learning how to market the NTFPs on the Internet is an option, but may entail higher shipping costs. Landowners should consider all options for selling their products including, farmer's markets or restaurants that focus on locally grown ingredients. The development phase should include a forest management plan that states the landowner's objectives and a resource inventory. Start-up costs should be analysed as specific equipment may be necessary to harvest or process the product, whereas other crops require minimal initial investment. Local incentives for sustainable forest management, as well as regulations and policies should be explored.

Benefits

- Extra income and diversification: Cultivation of specialty crops in a forest setting provides new sources of annual or periodic income before, or instead of, timber harvests.
- High value products: A wide variety of plants can be grown in a forest to produce natural or processed (value-added) edible, herbal, medicinal, decorative and craft products.
- Ecologically responsible: The intensively-managed cultivation of forest plants that are endangered in the wild due to over-harvesting is the responsible alternative to wildcrafting.

Variety of Non-timber Forest Products can be Grown

- Suitable species: Shade-loving plants that are naturally adapted to grow under forest conditions are candidates for forest farming cultivation.
- Research cultivation methods: Gather information about the biological and cultural requirements of understory plants in your forest. If possible, talk to other growers in your area. Research propagation methods and find sources of seed or cuttings.

Forest Management Aspects

- Forest management plan: Any forest farming enterprises should be integrated within your forest management plan. Determine any needed actions such as thinning, and any potential conflicts with silvicultural practices, e.g., herbicide applications.
- Silvicultural benefits: Management of the forest canopy to provide optimum light levels for understory crops, e.g. by thinning and pruning, also improves tree growth and wood quality.

Marketing Non-timber Products

- Inventory: The first step is to identify native plants growing in your forest that may have economic value, and to assess the soil and climatic conditions of the site. The understory plant community changes as the forest grows and with silvicultural activities. Determine all the possible uses of the plants and products that can be made from them.

- Market survey: Visit both retail outlet and wholesale buyers in your area to learn what forest-grown products they buy, including those produced from plants growing in your forest or which could be cultivated there. Learn about the buyer's specific requirements regarding quantities, sizes, seasons, prices, etc.
- Direct marketing: Retailing value-added products made from understory crops directly to buyers takes time but will give the best return to the landowner. Potential buyers include restaurants, ethnic markets, natural food stores and the public via the Internet.
- Wholesale: By harvesting and processing understory crops grown in your forest, they can be sold to wholesale buyers. Growers can benefit by forming marketing cooperatives to provide a larger, consistent supply of plant materials.
- Leasing: The simplest way to generate cash income may be to lease a portion of your forestland to others who will grow and harvest understory crops, although the return to the landowner will be less than other marketing methods. A clear lease contract is vital.

Forest Gardening

Forest gardening is a low-maintenance sustainable plant-based food production and agroforestry system based on woodland ecosystems, incorporating fruit and nut trees, shrubs, herbs, vines and perennial vegetables which have yields directly useful to humans. Making use of companion planting, these can be intermixed to grow in a succession of layers, to build a woodland habitat.

Forest gardening is a prehistoric method of securing food in tropical areas. In the 1980s, Robert Hart coined the term "forest gardening" after adapting the principles and applying them to temperate climates.

Forest gardens are probably the world's oldest form of land use and most resilient agroecosystem. They originated in prehistoric times along jungle-clad river banks and in the wet foothills of monsoon regions. In the gradual process of families improving their immediate environment, useful tree and vine species were identified, protected and improved whilst undesirable species were eliminated. Eventually superior foreign species were selected and incorporated into the gardens.

Robert Hart's forest garden in Shropshire.

Forest gardens are still common in the tropics and known by various names such as: home gardens, Zambia, Zimbabwe and Tanzania; Kandyan forest gardens in Sri Lanka; huertos familiares, the "family orchards" of Mexico. These are also called agroforests and, where the wood components are short-statured, the term shrub garden is employed. Forest gardens have been shown to be a significant source of income and food security for local populations.

Robert Hart adapted forest gardening for the United Kingdom's temperate climate during the 1980s. His theories were later developed by Martin Crawford from the Agroforestry Research Trust and various permaculturalists such as Graham Bell, Patrick Whitefield, Dave Jacke and Geoff Lawton.

Temperate Climates

Robert Hart, forest gardening pioneer.

Robert Hart coined the term "forest gardening" during the 1980s. Hart began farming at Wenlock Edge in Shropshire with the intention of providing a healthy and therapeutic environment for himself and his brother Lacon. Starting as relatively conventional smallholders, Hart soon discovered that maintaining large annual vegetable beds, rearing livestock and taking care of an orchard were tasks beyond their strength. However, a small bed of perennial vegetables and herbs he planted was looking after itself with little intervention.

Following Hart's adoption of a raw vegan diet for health and personal reasons, he replaced his farm animals with plants. The three main products from a forest garden are fruit, nuts and green leafy vegetables. He created a model forest garden from a 0.12 acre (500 m^2) orchard on his farm and intended naming his gardening method ecological horticulture or ecocultivation. Hart later dropped these terms once he became aware that agroforestry and forest gardens were already being used to describe similar systems in other parts of the world.

Seven-layer System

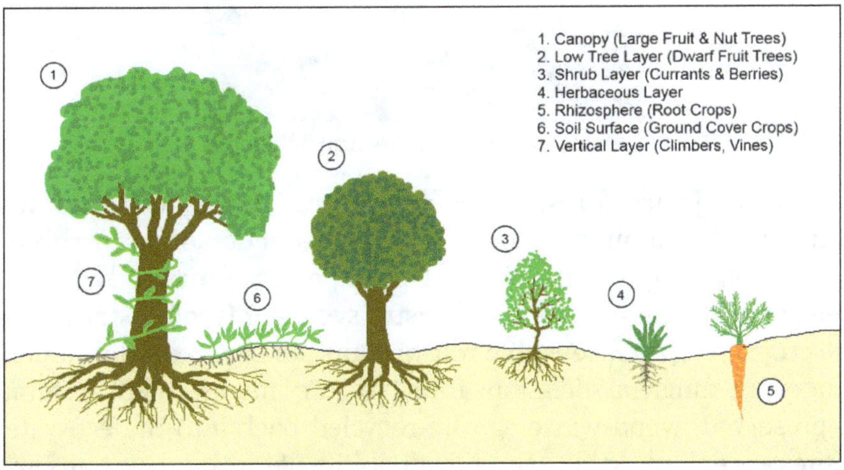

The seven layers of the forest garden.

Robert Hart pioneered a system based on the observation that the natural forest can be divided into distinct levels. He used intercropping to develop an existing small orchard of apples and pears into an edible polyculture landscape consisting of the following layers:

- 'Canopy layer' consisting of the original mature fruit trees.
- 'Low-tree layer' of smaller nut and fruit trees on dwarfing root stocks.
- 'Shrub layer' of fruit bushes such as currants and berries.
- 'Herbaceous layer' of perennial vegetables and herbs.
- 'Rhizosphere' or 'underground' dimension of plants grown for their roots and tubers.
- 'Ground cover layer' of edible plants that spread horizontally.

- 'Vertical layer' of vines and climbers.

A key component of the seven-layer system was the plants he selected. Most of the traditional vegetable crops grown today, such as carrots, are sun-loving plants not well selected for the more shady forest garden system.

Mycoforestry

Amanita species are ectomycorrhizal with many trees.

Mycoforestry is an ecological forest management system implemented to enhance forest ecosystems and plant communities through the introduction of mycorrhizal and saprotrophic fungi. Mycoforestry is considered a type of permaculture and can be implemented as a beneficial component of an agroforestry system. Mycoforestry can enhance the yields of tree crops and produce edible mushrooms, an economically valuable product. By integrating plant-fungal associations into a forestry management system, native forests can be preserved, wood waste can be recycled back into the ecosystem, planted restoration sites are enhanced, and the sustainability of forest ecosystems are improved. Mycoforestry is an alternative to the practice of clearcutting, which removes dead wood from forests, thereby diminishing nutrient availability and reducing soil depth.

Selection of Fungal Species

According to Paul Stamets, the first principle for the creation of a mycoforestry system is to utilize native fungal species. Implementing a mycoforestry system provides the potential of improving restoration efforts and the possibility of economic gain through mushroom cropping and harvesting. However to utilize native fungal flora, first the relationships between present fungal species and growth substrate, and habitat need to be studied.

A simple way to introduce a mycoforestry system and enhance out-plantings for crops and forest restoration sites is to "use mycorrhizal spore inoculum when replanting forest lands" For this process it is best to match native trees with native mycorrhizal fungi. This method keeps and will promote the functioning of the native ecosystem, and native biodiversity.

It is assumed in a functioning forest ecosystem an underground mycelial network persists even if no fruiting bodies are visible. A period of disappearance of mushrooms from an area should not cause alarm. In order to trigger the formation of fruiting bodies, many fungal species require specific environmental conditions. Most species of fungi do not fruit year round.

Mycoforestry is an emergent scientific field and practice. Until broadly standardized protocols are created and perfected, the collection of both current and historical ecological site conditions will improve the success of the project. Therefore, a survey of fungal relations at the site under both prime and poor conditions is beneficial to implementation of a mycoforestry system.

Saprotrophic Fungi

Edible oyster mushrooms (Pleurotus sp.) fruiting from a stump.

The second principle is to promote saprotrophic fungi in the environment. Saprophytic fungi are crucial to mycoforestry systems because these are the primary composers breaking down wood and returning nutrients to the soil for use by the rest of the forest ecosystem. This can be accomplished through inoculation of wood debris at site. Spored oils can be used in chainsaws when problematic or invasive hardwood requires felling. This method is a simple means to inoculate a tree. Additionally plug spawn can be implemented and injected into wood mass again prompting colonization by the selected fungus. Eventually repeated colonization efforts should not be necessary as many fungal life forms are strong and will spread and sustain in the soil on their own.

In management of the mycoforestry system it is important that dead wood be in contact with the ground. This allows fungus to reach up from the soil and decompose fallen wood releasing nutrients at a much quicker rate then if the wood is left standing. Additionally it is important to leave dead wood on site for decomposition back into the soil. This philosophy is similarly based to the fact that clear cutting of a forest reduces soil nutrients and thickness.

Beneficial Fungal Interactions

The third principal is to implement species known to benefit plant species. These are commonly mycorrhizal fungus that form long term associations with plants, often extending inside of plants roots acting as an additional root system providing for better absorption of nutrients and water.

Armillaria, a parasitic fungus.

Utilizing mushroom species that attract insects could be a useful source of fish food. This practice makes the mycoforestry a larger system. Unlike most agriculture systems it helps the environment in a number of ways. It ties all biological aspects of the environment together, creating sustainable living and food production as well as sustainable fisheries similar to the ancient Hawaiian Ahupua'a, which utilized sustainable all portions of the land for environmental and food security.

Additionally fungal species can be implemented that compete with disease causing agents like Armillaria root rots to provide long term protection of the forestry system.

Additionally, the implementation of an agroforestry system performs myco-remediation and myco-filtration activities cleaning up toxins and restoring the environment.

Riparian Buffer

A riparian buffer or stream buffer is a vegetated area (a "buffer strip") near a stream, usually forested, which helps shade and partially protect the stream from the impact of adjacent land uses. It plays a key role in increasing water quality in associated streams, rivers, and lakes, thus providing environmental benefits. With the decline of many aquatic ecosystems due to agriculture, riparian buffers have become a very common conservation practice aimed at increasing water quality and reducing pollution.

Benefits

Riparian buffers act to intercept sediment, nutrients, pesticides, and other materials in surface runoff and reduce nutrients and other pollutants in shallow subsurface water flow. They also serve to provide habitat and wildlife corridors in primarily agricultural areas. They can also be key in reducing erosion by providing stream bank stabilization.

Water Quality Benefits

- Intercepting sediments/nutrients: Key to counteract eutrophication in downstream lakes and ponds which can be detrimental to aquatic habitats because of large fish kills that occur upon large-scale eutrophication.

- Intercepting pesticides: Riparian buffers keep chemicals that can be harmful to aquatic life out of the water. Some pesticides can be especially harmful if they bio accumulate in the organism, with the chemicals reaching harmful levels once they are ready for human consumption.

- Bank stabilization: This is important because erosion can be a major problem in agricultural regions when cut (eroded) banks can take land out of production. Erosion can also lead to sedimentation and siltation of downstream lakes, ponds, and reservoirs. Siltation can greatly reduce the life span of reservoirs and the dams that create the reservoirs.

Habitat Benefits

- Provide habitat: Riparian buffers can act as crucial habitat for a large number of species, especially those who have lost habitat due to agricultural land being put into production.

- Increase biodiversity: By adding this vegetated area of land near a water source it becomes a prime location for species that may have left the area due to non-conservation land use to re-establish. With this re-establishment the number of native species and biodiversity in general can be increased.

- Buffers acting as corridors: Buffers also serve a major role in wildlife habitat. The habitat provided by the buffers also doubles as corridors for species that have had their habitat fragmented by various land uses.

- Shading water: The large trees in the first zone of the riparian buffer provide shade and therefore cooling for the water, increasing productivity and increasing habitat quality for aquatic species.

- Large woody debris: When branches and stumps fall into the stream from the riparian zone, more stream habitat features are created. Carbon is added as an energy source for biota in the stream.

Economic Benefits

- Increase land value: Often people who purchase land for recreational use are willing to pay more if there is more wooded area located on the land.

- Produce profitable alternative crops: Vegetation such as Black Walnut and Hazelnut, which can be profitably harvested, can be incorporated into the riparian buffer.

- Increase lease fees for hunting: The increased habitat means that the land will be more sought-after for hunting purposes.

Buffer Design

A riparian buffer is usually split into three different zones, each having its own specific purpose for filtering runoff and interacting with the adjacent aquatic system. Buffer design is a key element in the effectiveness of the buffer. It is generally recommended that native species be chosen to plant in these three zones, with the general width of the buffer being 50 feet (15 m) on each side of the stream.

Zone 1: This zone should function mainly to shade the water source and act as a bank stabilizer. The zone should include large native tree species that grow fast and can quickly act to perform these tasks. Although this is usually the smallest of the three zones and absorbs the fewest contaminants, most of the contaminants have been eliminated by Zone 2 and Zone 3.

Zone 2: Usually made up of native shrubs, this zone provides a habitat for wildlife, including nesting areas for bird species. This zone also acts to slow and absorb contaminants that Zone 3 has missed. The zone is an important transition between grassland and forest.

Zone 3: This zone is important as the first line of defense against contaminants. It consists mostly of native grasses and serves primarily to slow water runoff and begin to absorb contaminants before they reach the other zones. Although these grass strips should be one of the widest zones, they are also the easiest to install.

Streambed Zone: The streambed zone of the riparian area is linked closely to Zone 1. Zone 1 provides fallen limbs, trees, and tree roots that in turn slow water flow, reducing erosional processes associated with increased water flow and flooding. This woody debris also increases habitat and cover for various aquatic species

Managing Forests in Riparian Area

Logging is sometimes recommended as a management practice in riparian buffers, usually to provide economic incentive. However, some studies have shown that logging can harm wildlife populations, especially birds. A study by the University of Minnesota found that there was a correlation between the harvesting of timber in riparian buffers and a decline in bird populations. Therefore, logging is generally discouraged as an environmental practice, and left to be done in designated logging areas.

Effectiveness

Riparian buffers have undergone much scrutiny about their effectiveness, resulting in thorough testing and monitoring. A study, conducted over a nine-year period, monitored the amounts of fertilizers that reached the watershed from the source of the application. It found that these buffers removed at least 60% of the nitrogen in the runoff, and at least 65% of the phosphorus from the fertilizer application. The same study showed that the effectiveness of the Zone 3 was much greater than that of both Zone 1 and 2 at removing contaminants. But another recent study did not find efficiency (or a very limiting capacity) for reducing glyphosate and AMPA leaching to streams ; Spontaneous herbaceous vegetation RBS is as efficient as Salix plantations and measures of Glyphosate in runoff after a year, suggest an unexpected persistence and even a capacity of RBS to potentially favor glyphosate infiltration up to 70 cm depth in the soil.

Long-term Sustainability

After the initial installation of the riparian buffer, relatively little maintenance needs to be performed to keep the buffer in good condition. Once the trees and grasses reach maturity, they regenerate naturally and make a more effective buffer. The sustainability of the riparian buffer makes it extremely attractive to landowners, since they do relatively little work and still receive payments. Riparian buffers have the potential to be the most effective ways to protect aquatic biodiversity, water quality and manage water resources in developing countries that lack the funds to install water treatment and supply systems in midsize and small towns.

Riparian buffers are the natural vegetation from the edge of the stream bank out through the riparian zone. The vegetative zone serves as a buffer to pollutants entering a stream from runoff, controls erosion, and provides habitat and nutrient input into the stream. A relatively undisturbed riparian zone supports a robust stream system; narrow riparian zones occur when roads, parking lots, fields, lawns, bare soil, rocks, or buildings are near the stream bank. Residential developments, urban centers, golf courses, and rangeland are the common causes of anthropogenic degradation of the riparian zone.

Riparian buffers are the most valuable protection a stream system has against outside influences. In most cases healthy riparian directly reflects upon the condition of the stream unless the source of the insult is a specific pollutant. Enhancement of the riparian buffer by re-planting native grasses, forbs, shrubs and trees is the first step in the recovery of the stream back to a more natural condition.

- Provides organic material as food for invertebrate, fish and wildlife.
- Supplies large and small pieces of woody debris that provide habitat for fish, invertebrates and amphibians.
- Alters how sunlight reaches the stream and is an important temperature moderator.
- Stabilizes stream banks and reduces erosion.
- Filters sediment and materials from overland runoff and roots of many plants traps and holds the sediments.

- Absorbs nutrients from overland and sub-surface flows.
- Reduces the impacts of flooding through temporary storage, interception and slow releases from heavy rains.

References

- Agroforestry-and-its-benefits: en.reset.org, Retrieved 08 August, 2020
- Crawford, Martin 2010. Creating a Forest Garden: Working with Nature to Grow Edible Crops. Totnes: Green Books. ISBN 1-900322-62-5
- Forest-farming: fs.usda.gov, Retrieved 17 May, 2020
- Vaughan, R. C.; J. F. Munsell; J. L. Chamberlain (2013). "Opportunities for Enhancing Non-timber Forest Products Management in the United States". Journal of Forestry. 111 (1): 26–33. doi:10.5849/jof.10-106
- Forest-farming, what-is-agroforestry: aftaweb.org, Retrieved 17 June, 2020
- RiparianMagic: dep.wv.gov, Retrieved 20 May, 2020
- Stamets, Paul (2005). Mycelium running: how mushrooms can help save the world. Ten Speed Press. ISBN 1-58008-579-2

5
Silviculture

The art, science and practice of controlling the structure, growth and quality of forests for meeting the needs of timber production is termed as silviculture. Ecological thinning, pruning, clearcutting, coppicing, etc. fall under its domain. This chapter closely examines silviculture to provide an extensive understanding of the subject.

Silviculture is the art and science of managing forests for desired outcomes, such as wildlife habitat, aesthetics, and timber production. This is accomplished by applying different types of cuttings at different times in your forest's development. These are called silvicultural treatments, and longterm planned sequences of these are called silvicultural systems. These systems are named by the number of age classes of trees that will result. The number of age classes is meaningful because it tells us about structure. All silviculture systems are based on the biological requirements of the trees themselves, how they are applied depends on the specific desires and goals of the landowner. Even-aged systems create and maintain one age class. These stands usually have foliage in a single high layer when mature, with little to no vegetation in the understory. Two-age systems have two age classes arranged in a two-storied structure. Unevenaged systems are characterized by three or more age classes with foliage typically dispersed from ground level to tree tops.

Treatments (cuttings) are defined by their purpose and the time of their application. Usually a silvicultural treatment is planned with either one or both of the following objectives in mind:

- To improve the composition of the forest and increase the growth of the remaining trees (intermediate treatments).
- To facilitate the production of new trees within or in place of the old forest (regeneration treatments).

Intermediate treatments or thinnings are applied to immature stands to redistribute growing space to desired trees. Regeneration treatments are applied to mature stands in order to establish a new age class of trees on the site, or to release existing seedlings. Regeneration treatments include:

Even-aged or Two-aged Treatments

- Clearcut: Seedlings establish in the open after all trees removed. Effective for regenerating shade intolerant. species (for example aspen or birch).
- Seed tree: Seedlings establish in the open from seed provided by scattered large trees retained after the cut. This does not work for shallowrooted and/or shade–tolerant species (for example spruce or fir).
- Shelterwood: Seedlings establish under existing trees which are removed in 1 to 3 cuts at 5 to 10 year intervals. This is flexible and works for many species, including shade-tolerants (for example spruce and fir), and mid-tolerants (for example white pine) depending on the canopy cover.

Uneven-aged Treatments

Selection—seedlings establish or are released in small gaps resulting from the removal of single or small groups of trees. These cuts are repeated at 10 to 20 year intervals, with the whole canopy never being removed at once.

Harvests that remove the poorest trees, reduce competition around your most desired stems, and establish regeneration of desirable species will likely improve your forest and offer you a range of future management options.

In addition there are other terms which you may run across. Partial cutting, for example, refers to any type of harvest that removes some, but not all, of the trees from a stand. This term doesn't really tell you much about the type of harvest being conducted. The same is true of the phrase selective cutting; this just means that trees are selected for removal. They could be the worst trees, thus improving the stand, or the best trees, thus degrading your stand. Some types of cutting, called diameter-limit cutting or high-grading are exploitative. These cuts remove the biggest and/or best trees from your woodlot and reduce the potential for future volume and value growth. Silviculture refers to sound forestry that is applied to achieve a specific outcome or residual condition. The focus is not on what you are harvesting, but on what is being left behind. That is why diameter-limit cutting and high-grading are not silviculture. They are defined solely by what is cut, and what is left may or may not be acceptable. Whenever cutting is done on your land, whether by you or by a logger, you should consider what you want your forest to look like after the operation. Harvests that remove the poorest trees, reduce competition around your most desired stems, and/or establish regeneration of desirable species will likely improve your forest and offer you a range of future management options. Harvests that appropriately apply silvicultural treatments have the potential to improve the health, productivity and value of your woodlot. With the advice of a forester, the services of a skilled logger and a basic understanding of silviculture you will be well equipped to implement sound management practices on your woodlot.

Regeneration

Regeneration is basic to the continuation of forested, as well as to the afforestation of treeless land. Regeneration can take place through self-sown seed ("natural regeneration"), by artificially sown seed, or by planted seedlings. In whichever case, the performance of regeneration depends on its growth potential and the degree to which its environment allows the potential to be expressed. Seed, of course, is needed for all regeneration modes, both for natural or artificial sowing and for raising planting stock in a nursery.

Natural regeneration is a "human-assisted natural regeneration" means of establishing a forest age class from natural seeding or sprouting in an area after harvesting in that area through selection cutting, shelter (or seed-tree) harvest, soil preparation, or restricting the size of a clear-cut stand to secure natural regeneration from the surrounding trees.

The process of natural regeneration involves the renewal of forests by means of self-sown seeds, root suckers, or coppicing. In natural forests, conifers rely almost entirely on regeneration through seed. Most of the broadleaves, however, are able to regenerate by the means of emergence of shoots from stumps (coppice) and broken stems.

Seedbed Requirements

Any seed, self-sown or artificially applied, requires a seedbed suitable for securing germination.

In order to germinate, a seed requires suitable conditions of temperature, moisture, and aeration. For seeds of many species, light is also necessary, and facilitates the germination of seeds in other species, but spruces are not exacting in their light requirements, and will germinate without light. White spruce seed germinated at 35 °F (1.7 °C) and 40 °F (4.4 °C) after continuous stratification for one year or longer and developed radicles less than 6 cm (2.4 in) long in the cold room. When exposed to light, those germinants developed chlorophyll and were normally phototropic with continued elongation.

For survival in the short and medium terms, a germinant needs: a continuing supply of moisture; freedom from lethal temperature; enough light to generate sufficient photosynthate to support respiration and growth, but not enough to generate lethal stress in the seedling; freedom from browsers, tramplers, and pathogens; and a stable root system. Shade is very important to the survival of young seedlings. In the longer term, there must be an adequate supply of essential nutrients and an absence of smothering.

In undisturbed forest, decayed windfallen stemwood provides the most favorable seedbed for germination and survival, moisture supply being dependable, and the elevation of seedlings somewhat above the general level of the forest floor reduces the danger of smothering by leaves and snow-pressed minor vegetation; nor is such a microsite likely to be subject to flooding. Advantages conferred by those microsites include: more light, higher temperatures in the rooting zone, and better mycorrhizal development. In a survey in the Porcupine Hills, Manitoba, 90% of all spruce seedlings were rooted in rotten wood.

Mineral soil seedbeds are more receptive than the undisturbed forest floor, and are generally moister and more readily rewetted than the organic forest floor. However, exposed mineral soil, much more so than organic-surfaced soil, is subject to frost heaving and shrinkage during drought. The forces generated in soil by frost or drought are quite enough to break roots.

The range of microsites occurring on the forest floor can be broadened, and their frequency and distribution influenced by site preparation. Each microsite has its own microclimate. Microclimates near the ground are better characterized by vapour pressure deficit and net incident radiation, rather than the standard measurements of air temperature, precipitation, and wind pattern.

Aspect is an important component of microclimate, especially in relation to temperature and moisture regimes. Germination and seedling establishment of Engelmann spruce were much better on north than on south aspect seedbeds in the Fraser Experimental Forest, Colorado; the ratios of seeds to 5-year-old seedlings were determined as 32:1, 76:1, and 72:1 on north aspect bladed-shaded, bladed-unshaded, and undisturbed-shaded seedbeds, respectively. Clear-cut openings of 1.2 to 2.0 hectares (3.0 to 4.9 acres) adjacent to an adequate seed source, and not more than 6 tree-heights wide, could be expected to secure acceptable regeneration (4,900, 5-year-old trees per hectare), whereas on undisturbed-unshaded north aspects, and on all seedbed treatments tested on south aspects, seed to seedling ratios were so high that the restocking of any clearcut opening would be questionable.

At least seven variable factors may influence seed germination: seed characteristics, light, oxygen, soil reaction (pH), temperature, moisture, and seed enemies. Moisture and temperature are the most influential, and both are affected by exposure. The difficulty of securing natural regeneration of Norway spruce and Scots pine in northern Europe led to the adoption of various forms of reproduction cuttings that provided partial shade or protection to seedlings from hot sun and wind. The main objective of echeloned strips or border-cuttings with northeast exposure was to protect regeneration from overheating, and was originated in Germany and deployed successfully by A. Alarik in 1925 and others in Sweden. On south and west exposures, direct insolation and heat reflected from tree trunks often result in temperatures lethal to young seedlings, as well as desiccation of the surface soil, which inhibits germination. The sun is less injurious on eastern exposures because of the lower temperature in the early morning, related to higher humidity and presence of dew.

In 1993, Henry Baldwin, after noting that summer temperatures in North America are often higher than those in places where border-cuttings have been found useful, reported the results of a survey of regeneration in a stand of red spruce plus scattered white spruce that had been isolated by clearcutting on all sides, so furnishing an opportunity for observing regeneration on different exposures in this old-field stand at Dummer, New Hampshire. The regeneration included a surprisingly large number of balsam fir seedlings from the 5% stand component of that species. The maximum density of spruce regeneration, determined 4 rods (20 m) inside from the edge of the stand on a north 20°E exposure, was 600,000/ha, with almost 100,000 balsam fir seedlings.

A prepared seedbed remains receptive for a relatively short period, seldom as long as 5 years, sometimes as short as 3 years. Seedbed receptivity on moist, fertile sites decreases with particular rapidity, and especially on such sites, seedbed preparation should be scheduled to take advantage of good seed years. In poor seed years, site preparation can be carried out on mesic and drier sites with more chance of success, because of the generally longer receptivity of seedbeds there than those on moister sites. Although an indifferent seed year can suffice if seed distribution is good and environmental conditions favourable to seedling germination and survival, small amounts of seed are particularly vulnerable to depredation by small mammals. Considerable flexibility is possible in timing site preparation to coincide with cone crops. Treatment can be applied either before any logging takes place, between partial cuts, or after logging. In cut and leave strips, seedbed preparation can be carried out as a single operation, pre-scarifying the leave strips, post-scarifying the cut strips.

Broadcast burning is not recommended as a method of preparing sites for natural regeneration, as it rarely exposes enough mineral soil to be sufficiently receptive, and the charred organic surfaces are a poor seedbed for spruce. A charred surface may get too hot for good germination and may delay germination until fall, with subsequent overwinter mortality of unhardened seedlings. Piling and burning of logging slash, however, can leave suitable exposures of mineral soil.

Artificial Regeneration

With a view to reducing the time needed to produce planting stock, experiments were carried out with white spruce and three other coniferous species from Wisconsin seed in the longer, frost-free growing season in Florida, 125 vs. 265 days in central Wisconsin and northern Florida, respectively. As the species studied are adapted to long photoperiods, extended day lengths of 20 hours were applied in Florida. Other seedlings were grown under extended day length in Wisconsin and with natural day length in both areas. After two growing seasons, white spruce under long days in Florida were about the same as those in Wisconsin, but twice as tall as plants under natural Wisconsin photoperiods. Under natural days in Florida, with the short local photoperiod, white spruce was severely dwarfed and had a low rate of survival. Black spruce responded similarly. After two growing seasons, long day plants of all 4 species in Florida were well balanced, with good development of both roots and shoots, equaling or exceeding the minimum standards for 2+1 and 2+2 outplanting stock of Lake States species. Their survival when lifted in February and outplanted in Wisconsin equalled that of 2+2 Wisconsin-grown transplants. Artificial extension of the photoperiod in the northern Lake States greatly increased height increment of white and black spruces in the second growing season.

Optimum conditions for seedling growth have been determined for the production of containerized planting stock. Alternating day/night temperatures have been found more suitable than a constant temperature; at 400 lumens/m^2 light regime, a 28 °C/20 °C day/night temperatures have been recommended for white spruce. However, temperature optima are not necessarily the same at different ages and sizes. In 1984, R. Tinus investigated the effects of combinations of day and night temperature on height, caliper, and dry weight of 4 seed sources of Engelmann spruce. The 4 seed sources appeared to have very similar temperature requirements, with night optima about the same of slightly lower than daylight optima.

Tree provenance is important in artificial regeneration. Good provenance takes into account suitable tree genetics and a good environmental fit for planted/seeded trees in a forest stand. The wrong genotype can lead to failed regeneration, or poor trees that are prone to pathogens and undesired outcomes.

Artificial regeneration has been a more common method involving planting because it is more dependable than natural regeneration. Planting can involve using seedlings (from a nursery), unrooted cuttings, or seeds.

Whichever method is chosen it can be assisted by tending techniques also known as intermediate stand treatments.

The fundamental genetic consideration in artificial regeneration is that seed and planting stock must be adapted to the planting environment. Most commonly, the method of managing seed and stock deployment is through a system of defined seed zones, within which seed and stock can be moved without risk of climatic maladaptation. Ontario adopted a seed zone system in the 1970s based on G.A. Hills' 1952 site regions and provincial resource district boundaries, but Ontario's seed zones are now based on homogeneous climatic regions developed with the Ontario Climate Model. The regulations stipulate that source-identified seedlots may be either a general collection, when only the seed zone of origin is known, or a stand collection from a specific latitude and longitude. The movement of general-collection seed and stock across seed zone boundaries is prohibited, but the use of stand-collection seed and stock in another seed zone is acceptable when the Ontario Climate Model shows that the planting site and place of seed origin are climatically similar. The 12 seed zones for white spruce in Quebec are based mainly on ecological regions, with a few modifications for administrative convenience.

Seed quality varies with source. Seed orchards produce seed of the highest quality, then, in order of decreasing seed quality produced, seed production areas and seed collection areas follow, with controlled general collections and uncontrolled general collections producing the least characterized seed.

Seeds

Dewinging and Extraction

When seed is first separated from cones it is mixed with foreign matter, often 2 to 5 times the volume of the seed. The more or less firmly attached membranous wings on the seed must be detached before it is cleaned of foreign matter. The testa must not incur damage during the dewinging process. Two methods have been used, dry and wet. Dry seed may be rubbed gently through a sieve that has a mesh through which only seed without wings can pass. Large quantities of seed can be processed in dewinging machines, which use cylinders of heavy wire mesh and rapidly revolving stiff brushes within to remove the wings. In the wet process, seed with wings attached are spread out 10 cm to 15 cm deep on a tight floor and slightly moistened throughout; light leather flails are used to free seed from the wings. B. Wang described a unique wet dewinging procedure in 1973 using a cement mixer, used at the Petawawa tree seed processing facility. Wings of white and Norway spruce seed can be removed by dampening the seed slightly before it is run through a fanning mill for the last time. Any moistened seed must be dried before fermentation or moulding sets in.

Seed Viability

A fluorescein diacetate (FDA) biochemical viability test for several species of conifer seed, including white spruce, estimates the proportion of live seed (viability) in a seedlot, and hence the percentage germination of a seedlot. The accuracy of predicting percentage germination was within +/- 5 for most seedlots. White spruce seed can be tested for viability by an indirect method, such as the fluorescein diacetate (FDA) test or 'Ultra-sound'; or by the direct growth method of 'germination'. Samples of white spruce seed inspected in 1928 varied in viability from 50% to 100%, but averaged 93%. A 1915 inspection reported 97% viability for white spruce seed.

Germinative Testing

The results of a germination test are commonly expressed as germinative capacity or a germination percentage, which is the percentage of seeds that germinate during a period of time, ending when germination is practically complete. During extraction and processing, white spruce seeds gradually lost moisture, and total germination increased. Mittal reported that white spruce seed from Algonquin Park, Ontario, obtained the maximum rate (94% in 6 days) and 99% total germination in 21 days after 14-week pre-chilling. The pre-treatment of 1% sodium hypochlorite increased germinability.

Encouraged by Russian success in using ultrasonic waves to improve the germinative energy and percentage germination of seeds of agricultural crops, Timonin demonstrated benefits to white spruce germination after exposure of seeds to 1, 2, or 4 minutes of ultrasound generated by an M.S.E. ultrasonic disintegrator with a power consumption of 280 VA and power impact of 1.35 amperes. However, no seeds germinated after 6 minutes of exposure to ultrasound.

Seed Dormancy

Seed dormancy is a complex phenomenon and is not always consistent within species. Cold stratification of white spruce seed to break dormancy has been specified as a requirement, but Heit and Hellum regarded stratification as unnecessary. Cone handling and storage conditions affect dormancy in that cold, humid storage (5 °C, 75% to 95% relative humidity) of the cones prior to extraction seemingly eliminated dormancy by overcoming the need to stratify. Periods of cold, damp weather during the period of cone storage might provide natural cold (stratification) treatment. Once dormancy was removed in cone storage, subsequent kiln-drying and seed storage did not reactivate dormancy.

Haddon and Winston found a reduction in viability of stratified seeds after 2 years of storage and suggested that stress might have been caused by stratification, e.g., by changes in seed biochemistry, reduced embryo vigor, seed aging or actual damage to the embryo. They further questioned the quality of the 2-year-old seed even though high germination occurred in the samples that were not stratified.

Cold Stratification

Cold stratification is the term applied to the storing of seeds in (and, strictly, in layers with) a moist medium, often peat or sand, with a view to maintaining viability and overcoming dormancy. Cold stratification is the term applied to storage at near-freezing temperatures, even if no medium is used. A common method of cold stratification, is to soak seed in tap water for up to 24 h, superficially dry it, then store moist for some weeks or even months at temperatures just above freezing. Although Hellum found that cold stratification of an Alberta seed source led to irregular germination, with decreasing germination with increasing length of the stratification period, Hocking's paired test with stratified and nonstratified Alberta seed from several sources revealed no trends in response to stratification. Hocking suggested that seed maturity, handling, and storage needed to be controlled before the need for stratification could be determined. Later, Winston and Haddon found that the storage of white spruce cones for 4 weeks at 5 °C prior to extraction obviated the need for stratification.

Seed Ripeness

Seed maturity cannot be predicted accurately from cone flotation, cone moisture content, cone specific gravity; but the province of B.C. found embryo occupying 90% + of the corrosion cavity and mega gametophyte being firm and whitish in colour are the best predictors for white spruce in B.C., and Quebec can forecast seed maturity some weeks in advance by monitoring seed development in relation to heat-sums and the phenological progression of the inflorescence of fireweed (Epilobium angustifolium L.), an associated plant species. Cone collection earlier than one week before seed maturity would reduce seed germination and viability during storage. Four stages of maturation were determined by monitoring carbohydrates, polyols, organic acids, respiration, and metabolic activity. White spruce seeds require a 6-week post-harvest ripening period in the cones to obtain maximum germinability, however, based on cumulative degree-days, seed from the same trees and stand showed 2-week cone storage was sufficient.

Forest Tree Plantations

Plantation Establishment Criteria

Plantations may be considered successful when outplant performance satisfies certain criteria. The term "free growing" is applied in some jurisdictions. Ontario's "Free-to-Grow" (FTG) equivalent relates to a forest stand that meets a minimum stocking standard and height requirement, and is essentially free of competition from surrounding vegetation that might impede growth. The FTG concept was introduced with the advent of the Forest Management Agreement program in Ontario in 1980 and became applicable to all management units in 1986. Policy, procedures, and methodologies readily applicable by forest unit managers to assess the effectiveness of regeneration programs were still under development during the Class Environmental Assessment hearings.

In British Columbia, the Forest Practices Code governs performance criteria. To minimize the subjectivity of assessing deciduous competition as to whether or not a plantation is established, minimum specifications of number, health, height, and competition have been specified in British Columbia. However, minimum specifications are still subjectively set and may need to be fine-tuned in order to avoid unwarranted delay in according established status to a plantation. For example, a vigorous white spruce with a strong, multi-budded leading shoot and its crown fully exposed to light on 3 sides would not qualify as free-growing in the current British Columbia Code but would hardly warrant description as unestablished.

Tending

Tending is the term applied to pre-harvest silvicultural treatment of forest crop trees at any stage after initial planting or seeding. The treatment can be of the crop itself (e.g., spacing, pruning, thinning, and improvement cutting) or of competing vegetation (e.g., weeding, cleaning).

Planting

How many trees per unit area (spacing) that should be planted is not an easily answered question. Establishment density targets or regeneration standards have commonly been based on traditional practice, with the implicit aim of getting the stand quickly to the free-to-grow stage. Money is wasted if more trees are planted than are needed to achieve desired stocking rates, and the chance to establish other plantations is proportionately diminished. Ingress (natural regeneration) on a site is difficult to predict and often becomes surprisingly evident only some years after planting has been carried out. Early stand development after harvesting or other disturbance undoubtedly varies greatly among sites, each of which has its own peculiar characteristics.

For all practical purposes, the total volume produced by a stand on a given site is constant and optimum for a wide range of density or stocking. It can be decreased, but not increased, by altering the amount of growing stock to levels outside this range. Initial density affects stand development in that close spacing leads to full site utilization more quickly than wider spacing. Economic operability can be advanced by wide spacing even if total production is less than in closely spaced stands.

Beyond the establishment stage, the relationship of average tree size and stand density is very important. Various density-management diagrams conceptualizing the density-driven stand dynamics have been developed. Smith and Brand's diagram has mean tree volume on the vertical axis and the number of trees/ha on the horizontal axis: a stand can either have many little trees or a few big ones. The self-thinning line shows the largest number of trees of a given size/ha that can be carried at any given time. However, Willcocks and Bell caution against using such diagrams unless specific knowledge of the stand trajectory is known.

In the Lake States, plantations have been made with the spacing between trees varying from 3 by 3 to 10 by 10 feet (0.9 m by 0.9 m to 3.0 m by 3.0 m). Kittredge recommended that no fewer than 600 established trees per acre (1483/ha) be present during the early life of a plantation. To insure this, at least 800 trees per acre (1077/ha) should be planted where 85% survival may be expected, and at least 1200/ac (2970/ha) if only half of them can be expected to live. This translates into recommended spacings of 5 by 5 to 8 by 8 feet (1.5 m by 1.5 m to 2.4 m by 2.4 m) for plantings of conifers, including white spruce in the Lake States.

Enrichment Planting

A strategy for enhancing natural forests' economic value is to increase their concentration of economically important, indigenous tree species by planting seeds or seedlings for future harvest, which can be accomplished with enrichment planting (EP). This means increasing the planting density (i.e., the numbers of plants per hectare) in an already growing forest stand.

Release Treatments

- Weeding: A process of getting rid of saplings' or seedlings' competition by mowing, application of herbicide, or other method of removal from the surroundings.

- Cleaning: Release of select saplings from competition by overtopping trees of a comparable age. The treatment favours trees of a desired species and stem quality.
- Liberation cutting: A treatment that releases tree seedling or saplings by removing older overtopping trees.

Spacing

Over-crowded regeneration tends to stagnate. The problem is aggravated in species that have little self-pruning ability, such as white spruce. Spacing is a thinning (of natural regeneration), in which all trees other than those selected for retention at fixed intervals are cut. The term juvenile spacing is used when most or all of the cut trees are unmerchantable. Spacing can be used to obtain any of a wide range of forest management objectives, but it is especially undertaken to reduce density and control stocking in young stands and prevent stagnation, and to shorten the rotation, i.e., to speed the production of trees of a given size. Volume growth of individual trees and the merchantable growth of stands are increased. The primary rationale for spacing is that thinning is the projected decline in maximum allowable cut. And since wood will be concentrated on fewer, larger, and more uniform stems, operating and milling costs will be minimized.

Methods for spacing may be: manual, using various tools, including power saws, brush saws, and clippers; mechanical, using choppers and mulchers; chemical; or combinations of several methods. One treatment has had notable success in spacing massively overstocked (<100 000 stems/ha) natural regeneration of spruce and fir in Maine. Fitted to helicopter, the Thru-Valve boom emits herbicide spray droplets 1000 μm to 2000 μm in diameter at very low pressure. Swaths 1.2 m wide and leave strips 2.4 m wide were obtained with “knife-edge” precision when the herbicide was applied by helicopter flying at a height of 21 m at a speed of 40–48 km/h. It seems likely that no other method could be as cost-effective.

Twenty years after spacing to 2.5 × 2.5 m, 30-year-old mixed stands of balsam fir and white spruce in the Green River watershed, New Brunswick, averaged 156.9 m^3/ha.

A spacing study of 3 conifers (white spruce, red pine and jack pine) was established at Moodie, Manitoba, on flat, sandy, nutritionally poor soils with a fresh moisture regime. Twenty years after planting, red pine had the largest average dbh, 15% greater than jack pine, while white spruce dbh was less than half that of the pines. Crown width showed a gradual increase with spacing for all 3 conifers. Results to date were suggesting optimum spacings between 1.8 m and 2.4 m for both pines; white spruce was not recommended for planting on such sites.

Comparable data are generated by espacement trials, in which trees are planted at a range of densities. Spacings of 1.25 m, 1.50 m, 1.75 m, 2.00 m, 2.50 m, and 3.00 m on 4 site classes were used in the 1922 trial at Petawawa, Ontario. In the first of 34 old fields white spruce plantations used to investigate stand development in relation to spacing at Petawawa, Ontario, regular rows were planted at average spacings of from 4 × 4 to 7 × 7 feet (1.22 m × 1.22 m to 2.13 m × 2.13 m). Spacings up to 10 × 10 feet (3.05 m × 3.03 m) were subsequently included in the study. Yield tables based on 50 years of data showed:

- Except for merchantable volumes at age 20 and site classes 50 and 60, closer spacings gave greater standing volumes at all ages than did wider spacings, the relative difference decreasing with age.
- Merchantable volume as a proportion of total volume increases with age, and is greater at wider than at closer spacings.
- Current annual volume increment culminates sooner at closer than at wider spacings.

A smaller espacement trial, begun in 1951 near Thunder Bay, Ontario, included white spruce at spacings of 1.8 m, 2.7 m, and 3.6 m. At the closest spacing, mortality had begun at 37 years, but not at the wider spacings.

The oldest interior spruce espacement trial in British Columbia was established in 1959 near Houston in the Prince Rupert Forest Region. Spacings of 1.2 m, 2.7 m, 3.7 m, and 4.9 m were used, and trees were measured 6, 12, 16, 26, and 30 years after planting. At wide espacements, trees developed larger diameters, crowns, and branches, but basal area and total volume/ha were greatest in the closest espacement. In more recent trials in the Prince George Region of British Columbia and in Manitoba, planting density of white spruce had no effect on growth after up to 16 growing seasons, even at spacings as low as 1.2 m. The slowness of juvenile growth and of crown closure delays the response to intra-competition. Initially, close spacing might even provide a positive nurse effect to offset any negative response to competition.

Thinning

Thinning is an operation that artificially reduces the number of trees growing in a stand with the aim of hastening the development of the remainder. The goal of thinning is to control the amount and distribution of available growing space. By altering stand density, foresters can influence the growth, quality, and health of residual trees. It also provides an opportunity to capture mortality and cull the commercially less desirable, usually smaller and malformed, trees. Unlike regeneration treatments, thinnings are not intended to establish a new tree crop or create permanent canopy openings.

Thinning greatly influences the ecology and micro-meteorology of the stand, lowering the inter-tree competition for water. The removal of any tree from a stand has repercussions on the remaining trees both above-ground and below. Silvicultural thinning is a powerful tool that can be used to influence stand development, stand stability, and the characteristics of the harvestable products.

When considering intensive conifer plantations designed for maximum production, it is essential to remember that tending and thinning regimes and wind and snow damage are intimately related.

Previous studies have demonstrated that repeated thinnings over the course of a forest rotation increase carbon stores relative to stands that are clear-cut on short rotations and that the carbon benefits differ according to thinning method.

Precommercial Thinning

In the early development of forest stand, density of trees remain high and there is competition among trees for nutrients. When natural regeneration or artificial seeding has resulted in dense, overstocked young stands, natural thinning will in most cases eventually reduce stocking to more silviculturally desirable levels. But by the time some trees reach merchantable size, others will be overmature and defective, and others will still be unmerchantable. To reduce this unbalance and to obtain more economic returns, in the early stage, one kind of cleaning is done which is known as precommercial thinning. Generally, one or two times precommercial thinning is done to facilitate the growth of the tree The yield of merchantable wood can be greatly increased and the rotation shortened by precommercial thinning. Mechanical and chemical methods have been applied, but their costliness has militated against their ready adoption.

Pruning

Pruning, as a silvicultural practice, refers to the removal of the lower branches of the young trees (also giving the shape to the tree) so clear knot-free wood can subsequently grow over the branch stubs. Clear knot-free lumber has a higher value. Pruning has been extensively carried out in the Radiata pine plantations of New Zealand and Chile, however the development of Finger joint technology in the production of lumber and mouldings has led to many forestry companies reconsidering their pruning practices. “Brashing” is an alternative name for the same process. Pruning can be done to all trees, or more cost effectively to a limited number of trees. There are two types of pruning: natural or self-pruning and artificial pruning. Most cases of self-pruning happen when branches do not receive enough sunlight and die. Wind can also take part in natural pruning which can break branches. Artificial pruning is where people are paid to come and cut the branches. Or it can be natural, where trees are planted close enough that the effect is to cause self-pruning of low branches as energy is put into growing up for light reasons and not branchiness.

Stand Conversion

The term stand conversion refers to a change from one silvicultural system to another and includes species conversion, i.e., a change from one species (or set of species) to another. Such change can be effected intentionally by various silvicultural means, or incidentally by default e.g., when high-grading has removed the coniferous content from a mixedwood stand, which then becomes exclusively self-perpetuating aspen. In general, such sites as these are the most likely to be considered for conversion.

Growth and Yield

In discussing yields that might be expected from the Canadian spruce forests, Haddock noted that Wright's quotation of spruce yields in the British Isles of 220 cubic feet per acre (15.4 m^3/ha) per year and in Germany of 175 cubic feet per acre (12.25 m^3/ha) per year was misleading, at least if it was meant to imply that such yields might be approached in the Boreal Forest Region of Canada. Haddock thought that Wright's suggestion of 20 to 40 (average 30) cubic feet per acre (1.4 m^3/ha to 2.8 m^3/ha (average 2.1 m^3/ha) per year was more reasonable, but still somewhat optimistic.

The principal way forest resource managers influence growth and yield is to manipulate the mixture of species and number (density) and distribution (stocking) of individuals that form the canopy of the stand. Species composition of much of the boreal forest in North America already differs greatly from its pre-exploitation state. There is less spruce and more hardwoods in the second-growth forest than in the original forest; Hearnden calculated that the spruce cover type had declined from 18% to only 4% of the total forested area in Ontario. Mixedwood occupies a greater proportion of Ontario's second-growth forest (41%) than in the original (36%), but its component of white spruce is certainly much diminished.

Growth performance is certainly influenced by site conditions and thus by the kind and degree of site preparation in relation to the nature of the site. It is important to avoid the assumption that site preparation of a particular designation will have a particular silvicultural outcome. Scarification, for instance, not only covers a wide range of operations that scarify, but also any given way of scarifying can have significantly different results depending on site conditions at the time of treatment. In point of fact, the term is commonly misapplied. Scarification is defined as "Loosening the top soil of open areas, or breaking up the forest floor, in preparation for regenerating by direct seeding or natural seedfall", but the term is often misapplied to practices that include scalping, screefing, and blading, which pare off low and surface vegetation, together with most off its roots to expose a weed-free surface, generally in preparation for sowing or planting thereon.

Thus, it is not surprising that literature can be used to support the view that the growth of seedlings on scarified sites is much superior to that of growth on similar sites that have not been scarified, while other evidence supports the contrary view that scarification can reduce growth. Detrimental results can be expected from scarification that impoverishes the rooting zone or exacerbates edaphic or climatic constraints.

Burning site preparation has enhanced spruce seedling growth, but it must be supposed that burning could be detrimental if the nutrient capital is significantly depleted.

An obvious factor greatly influencing regeneration is competition from other vegetation. In a pure stand of Norway spruce, for instance, Roussel found the following relationships:

Percent cover (%)	Vegetation Description
Below 1	No vegetation
1-3	Moss carpet with a few fir seedlings
4-10	Herbaceous plants appear
10-25	Bramble, herbs, fairly vigorous spruce seedlings
>25	Herbs, brambles very dense, vigorous, no moss

A factor of some importance in solar radiation–reproduction relationships is excess heating of the soil surface by radiation. This is especially important for seedlings, such as spruce, whose first leaves do not shade the base of the stem at the soil surface. Surface temperatures in sandy soils on occasion reach lethal temperatures of 50 °C to 60 °C.

Common Methods of Harvesting

Silvicultural regeneration methods combine both the harvest of the timber on the stand and re-establishment of the forest. The proper practice of sustainable forestry should mitigate the potential negative impacts, but all harvest methods will have some impacts on the land and residual stand. The practice of sustainable forestry limits the impacts such that the values of the forest are maintained in perpetuity. Silvicultural prescriptions are specific solutions to a specific set of circumstances and management objectives. Following are some common methods:

Clearcut Harvesting

Conventional clearcut harvesting is relatively simple: all trees on a cutblock are felled and bunched with bunches aligned to the skidding direction, and a skidder then drags the bunches to the closest log deck. Feller-buncher operators concentrate on the width of the felled swath, the number of trees in a bunch, and the alignment of the bunch. Providing a perimeter boundary is felled during daylight, night-shift operations can continue without the danger of trespassing beyond the block. Productivity of equipment is maximized because units can work independently of one another.

Clearcutting

An even-aged regeneration method that can employ either natural or artificial regeneration. It involves the complete removal of the forest stand at one time. Clearcutting can be biologically appropriate with species that typically regenerate from stand replacing fires or other major disturbances, such as Lodgepole Pine (Pinus contorta). Alternatively, clearcutting can change the dominating species on a stand with the introduction of non-native and invasive species as was shown at the Blodgett Experimental Forest near Georgetown California. Additionally, clearcutting can prolong slash decomposition, expose soil to erosion, impact visual appeal of a landscape and remove essential wildlife habitat. It is particularly useful in regeneration of tree species such as Douglas-fir (Pseudotsuga menziesii) which is shade intolerant. In addition, the general public's distaste for even-aged silviculture, particularly clearcutting, is likely to result in a greater role for uneven-aged management on public lands as well. Across Europe, and in parts of North America, even-aged, production-orientated and intensively managed plantations are beginning to be regarded in the same way as old industrial complexes: something to abolish or convert to something else.

Clearcutting will impact many site factors important in their effect on regeneration, including air and soil temperatures. Kubin and Kemppainen for instance, measured temperatures in northern Finland from 1974 through 1985 in 3 clear-felled areas and in 3 neighbouring forest stands dominated by Norway spruce. Clear felling had no significant influence on air temperature at 2 m above the ground surface, but the daily air temperature maxima at 10 cm were greater in the clear-felled area than in the uncut forest, while the daily minima at 10 cm were lower. Night frosts were more common in the clear-felled area. Daily soil temperatures at 5 cm depth were 2 °C to 3 °C greater in the clear-felled area than in the uncut forest, and temperatures at depths of 50 cm and 100 cm were 3 °C to 5 °C greater. The differences between the clear-felled and uncut areas did not diminish during the 12 years following cutting.

Coppicing

A regeneration method which depends on the sprouting of cut trees. Most hardwoods, the coast redwood, and certain pines naturally sprout from stumps and can be managed through coppicing. Coppicing is generally used to produce fuelwood, pulpwood, and other products dependent on small trees. A close relative of coppicing is pollarding. Three systems of coppice woodland management are generally recognized: simple coppice, coppice with standards, and the coppice selection system.

- In Compound coppicing or coppicing with standards, some of the highest quality trees are retained for multiple rotations in order to obtain larger trees for different purposes.

Direct Seeding

Prochnau, 4 years after sowing, found that 14% of viable white spruce seed sown on mineral soil had produced surviving seedlings, at a seed: seedling ratio of 7.1:1. With Engelmann spruce, Smith and Clark obtained average 7th year seed: seedling ratios of 21:1 on scarified seedbeds on dry sites, 38:1 on moist sites, and 111:1 on litter seedbeds.

Group Selection

The group selection method is an uneven-aged regeneration method that can be used when mid-tolerant species regeneration is desired. The group selection method can still result in residual stand damage in dense stands, however directional falling can minimize the damage. Additionally, foresters can select across the range of diameter classes in the stand and maintain a mosaic of age and diameter classes.

Méthode Du Contrôle

Classical European silviculture achieved impressive results with systems such as Henri Biolley's méthode du contrôle in Switzerland, in which the number and size of trees harvested were determined by reference to data collected from every tree in every stand measured every 7 years.

While not designed to be applied to boreal mixedwoods, the méthode du contrôle is described briefly here to illustrate the degree of sophistication applied by some European foresters to the management of their forests. Development of management techniques that allowed for stand development to be monitored and guided into sustainable paths were in part a response to past experience, particularly in Central European countries, of the negative effects of pure, uniform stands with species often unsuited to the site, which greatly increased the risk of soil degradation and biotic diseases. Increased mortality and decreased increment generated widespread concern, especially after reinforcement by other environmental stresses.

More or less uneven-aged, mixed forests of preponderantly native species, on the other hand, treated along natural lines, have proved to be healthier and more resistant to all kinds of external dangers; and in the long run such stands are more productive and easier to protect.

However, irregular stands of this type are definitely more difficult to manage—new methods and techniques had to be sought particularly for the establishment of inventories, as well as increment control and yield regulation. In Germany, for instance, since the beginning of the nineteenth century under the influence of G.L. Hartig, yield regulation has been effected almost exclusively by allotment or formula methods based on the conception of the uniform normal forest with a regular succession of cutting areas.

In France, on the other hand, efforts were made to apply another kind of forest management, one that aimed to bring all parts of the forest to a state of highest productive capacity in perpetuity. In 1878, the French forester A. Gurnaud published a description of a méthode du contrôle for determining increment and yield. The method was based on the fact that through careful, selective harvesting, the productivity of the residual stand can be improved, because timber is removed as a cultural operation. In this method, the increment of stands is accurately determined periodically with the object of gradually converting the forest, through selective management and continuous experimentation, to a condition of equilibrium at maximum productive capacity.

Henri Biolley was the first to apply Gurnaud's inspired ideas to practical forestry. From 1890 on, he managed the forests of his Swiss district according to these principles, devoting himself for almost 50 years to the study of increment and a treatment of stands directed towards the highest production, and proving the practicability of the check method. In 1920, he published this study giving a theoretical basis of management of forests under the check method, describing the procedures to be applied in practice (which he partly developed and simplified), and evaluating the results.

Biolley's pioneering work formed the basis upon which most Swiss forest management practices were later developed, and his ideas have been generally accepted. Today, with the trend of intensifying forest management and productivity in most countries, the ideas and application of careful, continuous treatment of stands with the aid of the volume check method are meeting with ever-growing interest. In Britain and Ireland, for example, there is increased application of Continuous Cover Forestry principles to create permanently irregular structures in many woodlands.

Patch Cut

Row and Broadcast Seeding

Spot and row seeders use less seed that does broadcast ground or aerial seeding but may induce clumping. Row and spot seeding confer greater ability to control seed placement than does broadcast seeding. Also, only a small percentage of the total area needs to be treated.

In the aspen type of the Great Lakes region, direct sowing of the seed of conifers has usually failed. However, Gardner after trials in Yukon, which included broadcast seeding of white spruce seed at 2.24 kg/ha that secured 66.5% stocking in the Scarified Spring Broadcast treatment 3 years after seeding, concluded that the technique held "considerable promise".

Seed-tree

An even-aged regeneration method that retains widely spaced residual trees in order to provide uniform seed dispersal across a harvested area. In the seed-tree method, 2-12 seed trees per acre (5-30/ha) are left standing in order to regenerate the forest. They will be retained until regeneration has become established at which point they may be removed. It may not always be economically viable or biologically desirable to re-enter the stand to remove the remaining seed trees. Seed-tree cuts can also be viewed as a clearcut with natural regeneration and can also have all of the problems associated with clearcutting. This method is most suited for light-seeded species and those not prone to windthrow.

Selection Systems

Selection systems are appropriate where uneven stand structure is desired, particularly where the need to retain continuous cover forest for aesthetic or environmental reasons outweighs other management considerations. Selection logging has been suggested as being of greater utility than shelterwood systems in regenerating old-growth

Engelmann Spruce Sub-alpine Fir (ESSF) stands in southern British Columbia. In most areas, selection logging favours regeneration of fir more than the more light-demanding spruce. In some areas, selection logging can be expected to favour spruce over less tolerant hardwood species or lodge pole pine.

Shelter Spot Seeding

The use of shelters to improve germination and survival in spot seedings seeks to capture the benefits of greenhouse culture, albeit miniature. The Hakmet seed shelter, for instance, is a semi-transparent plastic cone 8 cm high, with openings of 7 cm diameter in the 7.5 cm diameter base and 17 mm diameter in the 24 mm diameter top. This miniature greenhouse increases air humidity, reduces soil desiccation, and raises air and soil temperatures to levels more favourable to germination and seedling growth than those offered by unprotected conditions. The shelter is designed to break down after a few years of exposure to ultraviolet radiation.

Seed shelters and spring sowing significantly improved stocking compared with bare spot seeding, but sheltering did not significantly improve growth. Stocking of bare seedspots was extremely low, possibly due to smothering of seedlings by abundant broadleaf and herbaceous litter, particularly that from aspen and red raspberry, and exacerbated by strong competition from graminoids and raspberry.

Cone shelters (Cerkon) usually produced greater survival than unsheltered seeding on scarified seedspots in trials of direct seeding techniques in interior Alaska, and funnel shelters (Cerbel) usually produced greater survival than unsheltered seeding on non-scarified seedspots. Both shelter types are manufactured by AB Cerbo in Trollhättan, Sweden. Both are made of light-degradable, white, opaque plastic, and are 8 cm high when installed.

White spruce seed was sown in Alaska on a burned site in summer 1984, and protected by white plastic cones on small spots scarified by hand, or by white funnels placed directly into the residual ash and organic material. A group of 6 ravens (Corvus corax) was observed in the area about 1 week after sowing was completed in mid-June. Damage averaged 68% with cones and 50% with funnels on an upland area, and 26% with funnels on a floodplain area. Damage by ravens was only 0.13% on unburned but otherwise similar areas.

In seeding trials in Manitoba between 1960 and 1966 aimed at converting aspen stands to spruce–aspen mixedwoods, 1961 scarification in the Duck Mountain Provincial Forest remained receptive to natural seeding for many years.

Shelterwood

In general terms, the shelterwood system is a series of partial cuts that removes the trees of an existing stand over several years and eventually culminates in a final cut that creates a new even-aged stand. It is an even-aged regeneration method that removes trees in a series of three harvests:

- Preparatory cut.
- Establishment cut.
- Removal cut.

The success of practising a shelterwood system is closely related to:

- The length of the regeneration period, i.e. the time from the shelterwood cutting to the date when a new generation of trees has been established.
- The quality of the new tree stands with respect to stand density and growth.
- The value increment of the shelter trees.

Information on the establishment, survival and growth of seedlings influenced by the cover of shelter trees, as well as on the growth of these trees, is needed as a basis for modelling the economic return of practising a shelterwood system. The method's objective is to establish new forest reproduction under the shelter of the retained trees. Unlike the seed-tree method, residual trees alter understory environmental conditions (i.e. sunlight, temperature, and moisture) that influence tree seedling growth. This method can also find a middle ground with the light ambiance by having less light accessible to competitors while still being able to provide enough light for tree regeneration. Hence, shelterwood methods are most often chosen for site types characterized by extreme conditions, in order to create a new tree generation within a reasonable time period. These conditions are valid foremost on level ground sites which are either dry and poor or moist and fertile.

Shelterwood Systems

Shelterwood systems involve 2, 3, or exceptionally more partial cuttings. A final cut is made once adequate natural regeneration has been obtained. The shelterwood system is most commonly applied as a 2-cut uniform shelterwood, first an initial regeneration (seed) cut, the second a final harvest cut. In stands less than 100 years old, a light preparatory cut can be useful. A series of intermediate cuts at intervals of 10–20 years has been recommended for intensively managed stands.

From operational or economic standpoints, however, there are disadvantages to the shelterwood system: harvesting costs are higher; trees left for deferred cutting may be damaged during the regeneration cut or related extraction operations; the increased risk of blowdown threatens the seed source; damage from bark beetles is likely to increase; regeneration may be damaged during the final cut and related extraction operations; the difficulty of any site preparation would be increased; and incidental damage to regeneration might be caused by any site preparation operations.

Single-tree Selection

The single-tree selection method is an uneven-aged regeneration method most suitable when shade tolerant species regeneration is desired. It is typical for older and diseased trees to be removed, thus thinning the stand and allowing for younger, healthy trees to grow. Single-tree selection can be very difficult to implement in dense or sensitive stands and residual stand damage can occur. This method is also disturbs the canopy layer the least out of all other methods.

Spot Seeding

Spot seeding was found to be the most economical and reliable of the direct seeding methods for converting aspen and paper birch to spruce and pine. In the Chippewa National Forest (Lake States), seed-spot sowing of 10 seeds each of white spruce and white pine under 40-year aspen after different degrees of cutting on gave second-season results clearly indicating the need to remove or disturb the forest floor to obtain germination of seeded white spruce and white pine.

Spot seeding of coniferous seed, including white spruce, has had occasional success, but several constraining factors commonly limit germination success: the drying out of the forest floor before the roots of germinants reach underlying moisture reserves; and, particularly under hardwoods, the smothering of small seedlings by snow-pressed leaf litter and lesser vegetation. Kittredge and Gervorkiantz determined that removal of the aspen forest floor increased germination percentage after the second season in seed spots of both white pine and white spruce, in 4 plots, from 2.5% to 5%, from 8% to 22%, from 1% to 9.5%, and from 0% to 15%.

Spot seeding requires less seed than broadcast seeding and tends to achieve more uniform spacing, albeit sometimes with clumping. The devices used in Ontario for manual spot seeding are the "oil can" seeder, seeding sticks, and shakers. The oil can is a container fitted with a long spout through which a predetermined number of seeds are released with each flick of the seeder.

Strip Cutting

Harvesting cutblocks where only a portion of the trees are to be removed is very different from clearcutting. First, trails must be located to provide access for the felling and skidding/forwarding equipment. These trails must be carefully located to ensure that the trees remaining meet the desired quality criteria and stocking density. Second, the equipment must not damage the residual stand. The further desiderata are outlined by Sauder.

The dearth of seed and a deficiency of receptive seedbeds were recognized as major reasons for the lack of success of clearcut harvesting. One remedy attempted in British Columbia and Alberta has been alternate strip cutting. The greater seed source from uncut trees between the cut strips, and the disturbance to the forest floor within the cut strips could be expected to increase the amount of natural regeneration. Trees were cut to a diameter limit in the cut strips, but large trees in the leave strips often proved too much of a temptation and were cut too, thus removing those trees that would otherwise have been the major source of seed.

An unfortunate consequence of strip thinning was the build-up of spruce beetle populations. Shaded slash from the initial cut, together with an increase in the number of wind thrown trees in the leave strips, provided conditions ideally suited to the beetle.

Underplanting

DeLong suggested underplanting 30- to 40-year-old aspen stands, on the basis of the success of natural spruce in regenerating under stands of such stands: "By planting, spacing can be controlled enabling easier protection of the spruce during stand entry for harvesting of the aspen overstorey".

Variable Retention

A harvesting and regeneration method which is a relatively new silvicultural system that retains forest structural elements (stumps, logs, snags, trees, understory species and undisturbed layers of forest floor) for at least one rotation in order to preserve environmental values associated with structurally complex forests.

"Uneven-aged and even-aged methods differ in the scale and intensity of disturbance. Uneven-aged methods maintain a mix of tree sizes or ages within a habitat patch by periodically harvesting individual or small groups of trees, Even-aged methods harvest most or all of the overstory and create a fairly uniform habitat patch dominated by trees of the same age". Even-aged management systems have been the prime methods to use when studying the effects on birds.

Mortality

A survey in 1955–56 to determine survival, development, and the reasons for success or failure of conifer pulpwood plantations (mainly of white spruce) in Ontario and Quebec up to 32 years old found that the bulk of the mortality occurred within the first 4 years of planting, unfavourable site and climate being the main causes of failure.

Advance Growth

Naturally regenerated trees in an understorey prior to harvesting constitute a classic case of good news and bad news. Understorey white spruce is of particular importance in mixedwoods dominated by aspen, as in the B15, B18a, and B19a Sections of Manitoba, and elsewhere. Until the latter part of the last century, white spruce understorey was mostly viewed as money in the bank on a long-term, low interest deposit, with final yield to be realized after slow natural succession, but the resource became increasingly threatened with the intensification of harvesting of aspen. White spruce plantations on mixedwood sites proved expensive, risky, and generally unsuccessful. These prompted efforts to see what might be done about growing aspen and white spruce on the same landbase by protecting existing white spruce advance growth, leaving a range of viable crop trees during the first cut and then harvesting both hardwoods and spruce in the final cut. Information about the understorey component is critical to spruce management planning. The ability of then current harvesting technology and crews employed to provide adequate protection for white spruce understories was questioned by Brace and Bella. Specialized equipment and training, perhaps with financial incentives, may be needed to develop procedures that would confer the degree of protection needed for the system to be feasible. Effective understorey management planning requires more than improved mixedwood inventory.

Avoidance of damage to the understorey will always be a desideratum. Sauder's paper on mixedwood harvesting describes studies designed to evaluate methods of reducing non-trivial damage to understorey residuals that would compromise their chance of becoming a future crop tree. Sauder concluded that:

- Operational measures that protected residual stems may not unduly increase costs.
- All felling, conifers and hardwoods, needs to be done in one operation to minimize the entry of the feller-buncher into the residual stand.
- Several operational procedures can reduce understorey damage, some of them without incurring extra costs.
- Successful harvesting of treatment blocks depends primarily on the intelligent location of skid trails and landings.

In summary, the key to protecting the white spruce understorey without sacrificing logging efficiency is a combination of good planning, good supervision, the use of appropriate equipment, and having conscientious, well-trained operators.Even the best plan will not reduce understorey damage unless its implementation is supervised.

New stands need to be established to provide for future supply of commercial white spruce from 150 000 ha of boreal mixedwoods in 4 of Rowe's regional Forest Sections straddling Alberta, Saskatchewan, and Manitoba, roughly from Peace River AB to Brandon MB. In the 1980s, with harvesting using conventional equipment and procedures, a dramatic increase in the demand for aspen posed a serious problem for the associated spruce understorey. Formerly, white spruce in the understories had developed to commercial size through natural succession under the protection of the hardwoods. Brace articulated a widespread concern: "The need for protection of spruce as a component of boreal mixedwoods goes beyond concern for the future commercial softwood timber supply. Concerns also include fisheries and wildlife habitat, aesthetics and recreation, a general dissatisfaction with cleacutting in mixedwoods and a strong interest in mixedwood perpetuation, as expressed recently in 41 public meetings on forestry development in northern Alberta".

On the basis of tests of 3 logging systems in Alberta, Brace affirmed that significant amounts of understorey can be retained using any of those systems provided that sufficient effort is directed towards protection. Potential benefits would include increased short-term softwood timber supply, improved wildlife habitat and cutblock aesthetics, as well as reduced public criticism of previous logging practices. Stewart developed statistical models to predict the natural establishment and height growth of understorey white spruce in the boreal mixedwood forest in Alberta using data from 148 permanent sample plots and supplementary information about height growth of white spruce regeneration and the amount and type of available substrate. A discriminant model correctly classified 73% of the sites as to presence or absence of a white spruce understorey, based on the amount of spruce basal area, rotten wood, ecological nutrient regime, soil clay fraction, and elevation, although it explained only 30% of the variation in the data. On sites with a white spruce understorey, a regression model related the abundance of regeneration to rotten wood cover, spruce basal area, pine basal area, soil clay fraction, and grass cover ($R^2 = 0.36$). About half the seedlings surveyed grew on rotten wood, and only 3% on mineral soil, and seedlings were 10 times more likely to have established on these substrates than on litter. Exposed mineral soil covered only 0.3% of the observed transect area.

Advance Growth Management

Advance growth management, i.e., the use of suppressed understorey trees, can reduce reforestation costs, shorten rotations, avoid denuding the site of trees, and also reduce adverse impacts on aesthetic, wildlife, and watershed values. To be of value, advance growth must have acceptable species composition and distribution, have potential for growth following release, and not be vulnerable to excessive damage from logging.

The age of advance growth is difficult to estimate from its size, as white that appears to be 2 to 3 year-old may well be more than 20 years old. However, age does not seem to determine the ability of advance growth of spruce to respond to release, and trees older than 100 years have shown rapid rates of growth after release. Nor is there a clear relationship between the size of advance growth and its growth rate when released.

Where advance growth consists of both spruce and fir, the latter is apt to respond to release more quickly than the former, whereas spruce does respond. If the ratio of fir to spruce is large, however, the greater responsiveness to release of fir may subject the spruce to competition severe enough to negate much of the effect of release treatment. Even temporary relief from shrub competition has increased height growth rates of white spruce in north-western New Brunswick, enabling the spruce to overtop the shrubs.

Site Preparation

Site preparation is any of various treatments applied to a site in order to ready it for seeding or planting. The purpose is to facilitate the regeneration of that site by the chosen method. Site preparation may be designed to achieve, singly or in any combination: improved access, by reducing or rearranging slash, and amelioration of adverse forest floor, soil, vegetation, or other biotic factors. Site preparation is undertaken to ameliorate one or more constraints that would otherwise be likely to thwart the objectives of management. A valuable bibliography on the effects of soil temperature and site preparation on subalpine and boreal tree species has been prepared by McKinnon

Site preparation is the work that is done before a forest area is regenerated. Some types of site preparation are burning.

Burning

Broadcast burning is commonly used to prepare clearcut sites for planting, e.g., in central British Columbia, and in the temperate region of North America generally.

Prescribed burning is carried out primarily for slash hazard reduction and to improve site conditions for regeneration; all or some of the following benefits may accrue:

- Reduction of logging slash, plant competition, and humus prior to direct seeding, planting, scarifying or in anticipation of natural seeding in partially cut stands or in connection with seed-tree systems.
- Reduction or elimination of unwanted forest covers prior to planting or seeding or prior to preliminary scarification thereto.
- Reduction of humus on cold, moist sites to favour regeneration.
- Reduction or elimination of slash, grass, or brush fuels from strategic areas around forested land to reduce the chances of damage by wildfire.

Prescribed burning for preparing sites for direct seeding was tried on a few occasions in Ontario, but none of the burns was hot enough to produce a seedbed that was adequate without supplementary mechanical site preparation.

Changes in soil chemical properties associated with burning include significantly increased pH, which Macadam in the Sub-boreal Spruce Zone of central British Columbia found persisting more than a year after the burn. Average fuel consumption was 20 to 24 t/ha and the forest floor depth was reduced by 28% to 36%. The increases correlated well with the amounts of slash (both total and ≥7 cm diameter) consumed. The change in pH depends on the severity of the burn and the amount consumed; the increase can be as much as 2 units, a 100-fold change. Deficiencies of copper and iron in the foliage of white spruce on burned clearcuts in central British Columbia might be attributable to elevated pH levels.

Even a broadcast slash fire in a clearcut does not give a uniform burn over the whole area. Tarrant, for instance, found only 4% of a 140-ha slash burn had burned severely, 47% had burned lightly, and 49% was unburned. Burning after windrowing obviously accentuates the subsequent heterogeneity.

Marked increases in exchangeable calcium also correlated with the amount of slash at least 7 cm in diameter consumed. Phosphorus availability also increased, both in the forest floor and in the 0 cm to 15 cm mineral soil layer, and the increase was still evident, albeit somewhat diminished, 21 months after burning. However, in another study in the same Sub-boreal Spruce Zone found that although it increased immediately after the burn, phosphorus availability had dropped to below pre-burn levels within 9 months.

Nitrogen will be lost from the site by burning, though concentrations in remaining forest floor were found by Macadam to have increased in 2 of 6 plots, the others showing decreases. Nutrient losses may be outweighed, at least in the short term, by improved soil microclimate through the reduced thickness of forest floor where low soil temperatures are a limiting factor.

The Picea/Abies forests of the Alberta foothills are often characterized by deep accumulations of organic matter on the soil surface and cold soil temperatures, both of which make reforestation difficult and result in a general deterioration in site productivity; Endean and Johnstone describe experiments to test prescribed burning as a means of seedbed preparation and site amelioration on representative clear-felled Picea/Abies areas. Results showed that, in general, prescribed burning did not reduce organic layers satisfactorily, nor did it increase soil temperature, on the sites tested. Increases in seedling establishment, survival, and growth on the burned sites were probably the result of slight reductions in the depth of the organic layer, minor increases in soil temperature, and marked improvements in the efficiency of the planting crews. Results also suggested that the process of site deterioration has not been reversed by the burning treatments applied.

Ameliorative Intervention

Slash weight (the oven-dry weight of the entire crown and that portion of the stem < 4 inches in diameter) and size distribution are major factors influencing the forest fire hazard on harvested sites. Forest managers interested in the application of prescribed burning for hazard reduction and silviculture, were shown a method for quantifying the slash load by Kiil. In west-central Alberta, he felled, measured, and weighed 60 white spruce, graphed (a) slash weight per merchantable unit volume against diameter at breast height (dbh), and (b) weight of fine slash (<1.27 cm) also against dbh, and produced a table of slash weight and size distribution on one acre of a hypothetical stand of white spruce. When the diameter distribution of a stand is unknown, an estimate of slash weight and size distribution can be obtained from average stand diameter, number of trees per unit area, and merchantable cubic foot volume. The sample trees in Kiil's study had full symmetrical crowns. Densely growing trees with short and often irregular crowns would probably be overestimated; open-grown trees with long crowns would probably be underestimated.

The need to provide shade for young outplants of Engelmann spruce in the high Rocky Mountains is emphasized by the U.S. Forest Service. Acceptable planting spots are defined as microsites on the north and east sides of down logs, stumps, or slash, and lying in the shadow cast by such material. Where the objectives of management specify more uniform spacing, or higher densities, than obtainable from an existing distribution of shade-providing material, redistribution or importing of such material has been undertaken.

Access

Site preparation on some sites might be done simply to facilitate access by planters, or to improve access and increase the number or distribution of microsites suitable for planting or seeding.

Wang determined field performance of white and black spruces 8 and 9 years after outplanting on boreal mixedwood sites following site preparation (Donaren disc trenching versus no trenching) in 2 plantation types (open versus sheltered) in southeastern Manitoba. Donaren trenching slightly reduced the mortality of black spruce but significantly increased the mortality of white spruce. Significant difference in height was found between open and sheltered plantations for black spruce but not for white spruce, and root collar diameter in sheltered plantations was significantly larger than in open plantations for black spruce but not for white spruce. Black spruce open plantation had significantly smaller volume (97 cm^3) compared with black spruce sheltered (210 cm^3), as well as white spruce open (175 cm^3) and sheltered (229 cm^3) plantations. White spruce open plantations also had smaller volume than white spruce sheltered plantations. For transplant stock, strip plantations had a significantly higher volume (329 cm^3) than open plantations (204 cm^3). Wang recommended that sheltered plantation site preparation should be used.

Mechanical

Up to 1970, no "sophisticated" site preparation equipment had become operational in Ontario, but the need for more efficacious and versatile equipment was increasingly recognized. By this time, improvements were being made to equipment originally developed by field staff, and field testing of equipment from other sources was increasing.

According to J. Hall, in Ontario at least, the most widely used site preparation technique was post-harvest mechanical scarification by equipment front-mounted on a bulldozer (blade, rake, V-plow, or teeth), or dragged behind a tractor (Imsett or S.F.I. scarifier, or rolling chopper). Drag type units designed and constructed by Ontario's Department of Lands and Forests used anchor chain or tractor pads separately or in combination, or were finned steel drums or barrels of various sizes and used in sets alone or combined with tractor pad or anchor chain units.

J. Hall's report on the state of site preparation in Ontario noted that blades and rakes were found to be well suited to post-cut scarification in tolerant hardwood stands for natural regeneration of yellow birch. Plows were most effective for treating dense brush prior to planting, often in conjunction with a planting machine. Scarifying teeth, e.g., Young's teeth, were sometimes used to prepare sites for planting, but their most effective use was found to be preparing sites for seeding, particularly in backlog areas carrying light brush and dense herbaceous growth. Rolling choppers found application in treating heavy brush but could be used only on stone-free soils. Finned drums were commonly used on jack pine–spruce cutovers on fresh brushy sites with a deep duff layer and heavy slash, and they needed to be teamed with a tractor pad unit to secure good distribution of the slash. The S.F.I. scarifier, after strengthening, had been "quite successful" for 2 years, promising trials were under way with the cone scarifier and barrel ring scarifier, and development had begun on a new flail scarifier for use on sites with shallow, rocky soils. Recognition of the need to become more effective and efficient in site preparation led the Ontario Department of Lands and Forests to adopt the policy of seeking and obtaining for field testing new equipment from Scandinavia and elsewhere that seemed to hold promise for Ontario conditions, primarily in the north. Thus, testing was begun of the Brackekultivator from Sweden and the Vako-Visko rotary furrower from Finland.

Mounding

Site preparation treatments that create raised planting spots have commonly improved outplant performance on sites subject to low soil temperature and excess soil moisture. Mounding can certainly have a big influence on soil temperature. Draper et al., for instance, documented this as well as the effect it had on root growth of outplants.

The mounds warmed up quickest, and at soil depths of 0.5 cm and 10 cm averaged 10 and 7 °C higher, respectively, than in the control. On sunny days, daytime surface temperature maxima on the mound and organic mat reached 25 °C to 60 °C, depending on soil wetness and shading. Mounds reached mean soil temperatures of 10 °C at 10 cm depth 5 days after planting, but the control did not reach that temperature until 58 days after planting. During the first growing season, mounds had 3 times as many days with a mean soil temperature greater than 10 °C than did the control microsites.

Draper et al.'s mounds received 5 times the amount of photosynthetically active radiation (PAR) summed over all sampled microsites throughout the first growing season; the control treatment consistently received about 14% of daily background PAR, while mounds received over 70%. By November, fall frosts had reduced shading, eliminating the differential. Quite apart from its effect on temperature, incident radiation is also important photosynthetically. The average control microsite was exposed to levels of light above the compensation point for only 3 hours, i.e., one-quarter of the daily light period, whereas mounds received light above the compensation point for 11 hours, i.e., 86% of the same daily period. Assuming that incident light in the 100-600 $\mu Em^{-2}s^{-1}$ intensity range is the most important for photosynthesis, the mounds received over 4 times the total daily light energy that reached the control microsites.

Orientation of Linear Site Preparation

With linear site preparation, orientation is sometimes dictated by topography or other considerations, but where the orientation can be chosen, it can make a significant difference. A disk-trenching experiment in the Sub-boreal Spruce Zone in interior British Columbia investigated the effect on growth of young outplants (lodgepole pine) in 13 microsite planting positions: berm, hinge, and trench; in north, south, east, and west aspects, as well as in untreated locations between the furrows. Tenth-year stem volumes of trees on south, east, and west-facing microsites were significantly greater than those of trees on north-facing and untreated microsites. However, planting spot selection was seen to be more important overall than trench orientation.

The N–S strips accumulated more snow, but the snow melted faster than on E–W strips in the first year after felling. Snow-melt was faster on strips near the centre of the strip-felled area than on border strips adjoining the intact stand. The strips, 50 feet (15.24 m) wide, alternating with uncut strips 16 feet (4.88 m) wide, were felled in a Pinus resinosa stand, aged 90 to 100 years.

Silvicultural System

A silvicultural system is a planned process whereby a stand is tended, harvested, and reestablished. It includes every management activity that occurs on that stand over an entire rotation. This is a useful scale for thinking about silviculture, both in spatial (one stand) and temporal (one rotation) terms.

While silviculture is typically practiced at the stand scale, forest management often occurs over multiple stands at the same time within a single land ownership. Silvicultural plans may be written for either shorter or longer time periods, depending on landowner objectives.

Silvicultural systems are commonly categorized based on their age class structure. Age class structure describes how many cohorts can be found in a forest.

- Even Age: All the trees are the same age. There is one cohort.
- Two Age: There are two distinct age classes of trees, one older, and one younger. There are two cohorts.

- Uneven age: There are three or more age classes of trees. There are at least three cohorts.

Management activities, or treatments, that together compose a silvicultural system can be broken down into several categories. These are regeneration treatments, stand establishment treatments, and intermediate treatments.

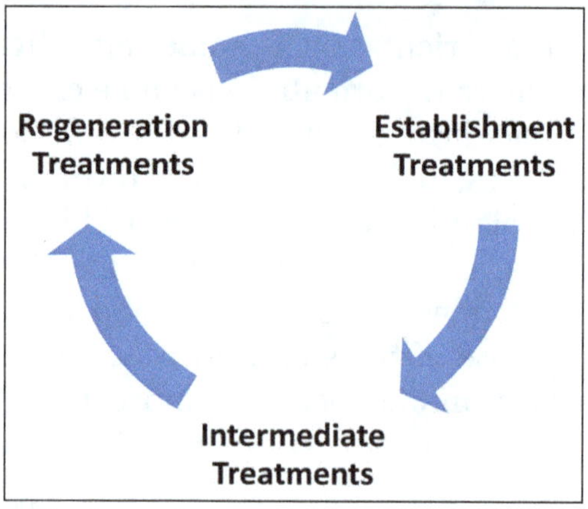

The three phases of a silvicultural system.

Regeneration Treatments

Regeneration treatments are harvest techniques that end one rotation and allow a new rotation to begin. A new rotation begins when the entire stand is established with one or more new cohorts. These are thus essentially harvest techniques. We call these harvest techniques regeneration treatments because the silviculturist's primary objective in harvesting a forest is to successfully establish the next rotation, which regenerates the forest. Regeneration methods are also categorized based on the age class structure of the newly established forest. This text will focus on eight distinct regeneration treatments:

- Even Age:
    - Clearcut.
    - Seed-Tree.
    - Shelterwood.
- Two Age:
    - With Reserves
    - Plus Deferment.
- Uneven Age:
    - Patch Selection.

- Group Selection.
- Single Tree Selection.

It is important to note that silvicultural systems are most commonly named after the regeneration method. This can cause confusion if you are uncertain whether a regeneration method alone is being referenced, or a whole silvicultural system that includes the regeneration method, establishment treatments, and intermediate treatments.

These regeneration treatments (and the silvicultural systems named after them) vary based on the age class structures they create, and on the amount of light, water, and nutrients that they make available to newly establishing cohorts.

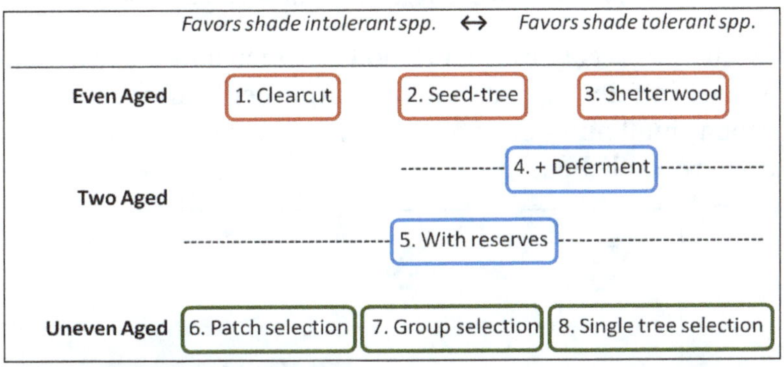

Eight methods of regeneration sorted by the age class structure they create and the shade tolerance of the species that they favor.

Even Aged Regeneration Treatments

Clearcut: A method of regenerating an even aged stand in which a new age class develops in a fully exposed microclimate after removal, in a single cutting, of all trees in the previous stand. Regeneration is from natural seeding, direct seeding, planted seedlings, and/or advance reproduction. The management unit or stand in which regeneration, growth, and yield are regulated consists of the individual clearcut stand.

A 120 acre clearcut on a xeric ridgetop in East Texas prior to site preparation and replanting. The previous and next rotations are both loblolly pine plantations. Notice that there is little slash or debris apparent on the site, a common feature of clearcut plantations of coniferous species.

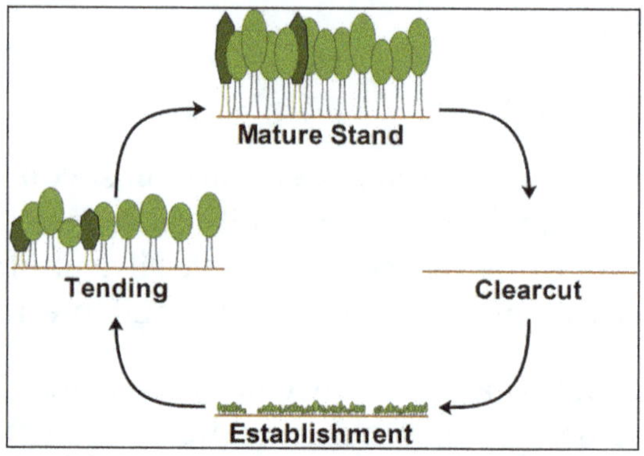

An example of the clearcut regeneration method. Note that this is a one-harvest system that regenerates an even-aged stand.

Seed-Tree: An even aged regeneration method in which a new age class develops from seeds that germinate in fully exposed microenvironments after removal of all the previous stand except a small number of trees left to provide seed. Seed trees are removed after regeneration is established.

A slash pine stand, with half the original stand still intact (right side) and the other half following the seed-tree cut (left side). Once the new cohort is established, the seed-trees will be removed.

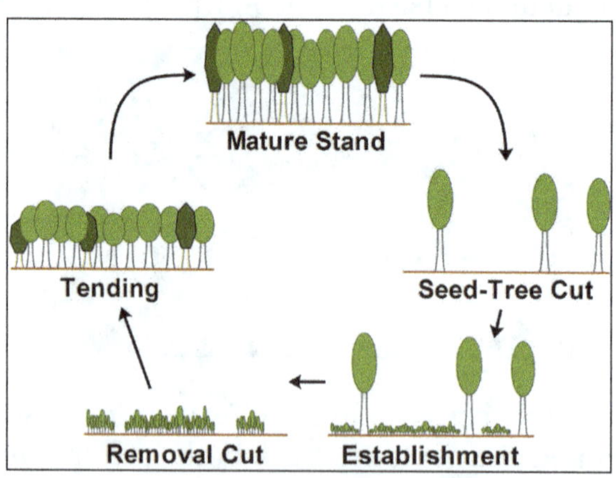

An example of the seed-tree regeneration method. Note that this is a two-harvest system that regenerates an even-aged stand.

Shelterwood: A method of regenerating an even aged stand in which a new age class develops beneath the moderated microenvironment provided by the residual trees. The sequence of treatments can include three distinct types of cuttings:

- An optional preparatory cut to enhance conditions for seed production.
- An establishment cut to prepare the seed bed and to create a new age class.
- A removal cut to release established regeneration from competition with the overwood.

Two Aged Regeneration Treatments

Plus Deferment: A two aged regeneration method in which a new age class develops from seeds that germinate under overwood that is not removed. This regeneration method modifies the seed-tree or shelterwood methods by skipping or indefinitely deferring the overwood removal or seed-tree removal cuts. The two cohorts are thus the newly established cohort, and the deferred seed trees or overwood. This is alternately known as the leave tree, reserve tree, or irregular shelterwood method.

A large eastern cottonwood apparent in a sugarberry, elm, cottonwood, sycamore stand regenerated with a deferment system in the US South.

An example of the shelterwood plus deferment regeneration method. Note that this is a one-harvest system that regenerates a two-aged stand. This is a modification of the shelterwood regeneration method in that the overwood is not removed in a second harvest, called the removal cut.

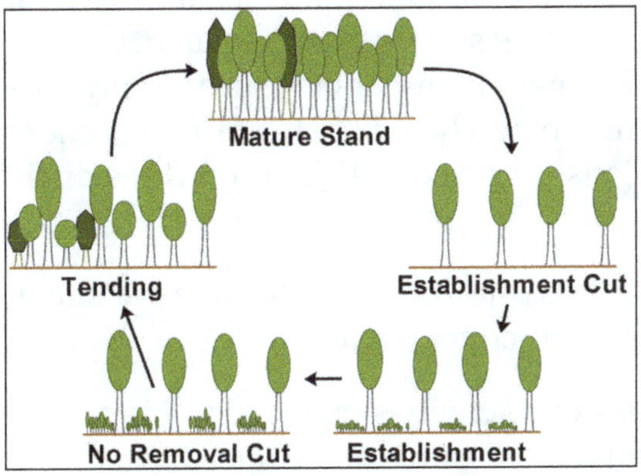

With Reserves: A two aged regeneration method in which a new age class develops from natural seeding, direct seeding, planted seedlings, and/or advance reproduction. This method is a modification of any of the three even aged systems that involves the retention of live or dead trees that are called reserves. The reserves can be in one or more groups, or may be dispersed throughout the stand.

Uneven Aged Regeneration Treatments

- Patch Selection: A method of regenerating uneven aged stands in which trees are removed, and new age classes are established, in large groups. The minimum width of groups is greater than twice the height of the mature trees, with large openings providing conditions suitable for intermediate to very intolerant regeneration. In the patch selection method, the management unit or stand in which regeneration, growth, and yield are regulated consists of a landscape containing an aggregation of patches.
- Group Selection: A method of regenerating uneven aged stands in which trees are removed, and new age classes are established, in small groups. The maximum width of groups is less than twice the height of the mature trees, with small openings providing microenvironment suitable for tolerant regeneration and the larger openings providing conditions suitable for more intolerant regeneration. In the group selection method, the management unit or stand in which regeneration, growth, and yield are regulated consists of a landscape containing an aggregation of groups.
- Single Tree Selection: A method of creating new age classes in uneven aged stands in which individual trees of all size classes are removed more-or-less uniformly throughout the stand to achieve desired stand structural characteristics.

Stand Establishment Treatments

Stand establishment treatments are any management activities that follow regeneration treatments and promote or facilitate the growth of new cohorts. Many traditional silvicultural texts consider regeneration treatments and stand establishment treatments as the same. This is for good reason, as often regeneration treatments will require no subsequent stand establishment treatments to successfully create a new stand. However, we will consider them separately to better inform our understanding of intensively managed forest plantations. This text will focus on three distinct stand establishment treatments:

- Site Preparation: Activities intended to improve site access, increase survival and early growth, and reduce competition.
- Natural Regeneration: Activities intended to foster the establishment of a new cohort from natural seed source or advanced regeneration.

- Artificial Regeneration: Activities including tree improvement, direct seeding, and planting.

Intermediate Treatments

Intermediate treatments are any management activities that take place between the time of stand establishment and the next regeneration treatment. Intermediate treatments are intended to affect stand growth rates, composition, and product class distribution. This text will focus on four distinct intermediate treatments:

- Competition Control: Application of chemical or mechanical methods to allocate a greater percentage of site resources to crop trees.
- Nutrition Management: Addition or manipulation of materials that affect the availability of mineral nutrients to crop trees.
- Pruning: Removing branches from crop trees to improve stem quality and product class.
- Thinning: Removing trees from a stand to reallocate site resources to crop trees and allow them to more rapidly grow into higher valued product classes.

Plant Nursery

Plants in a nursery.

A nursery is a place where plants are propagated and grown to a desired age. They include retail nurseries which sell to the general public, wholesale nurseries which sell only to businesses such as other nurseries and to commercial gardeners, and private nurseries which supply the needs of institutions or private estates.

Nurseries may supply plants for gardens, for agriculture, for forestry and for conservation biology.

Some of them specialize in one phase of the process: propagation, growing out, or retail sale; or in one type of plant: e.g., groundcovers, shade plants, or rock garden plants. Some produce bulk stock, whether seedlings or grafted, of particular varieties for purposes such as fruit trees for orchards, or timber trees for forestry. Some produce stock seasonally, ready in springtime for export to colder regions where propagation could not have been started so early, or to regions where seasonal pests prevent profitable growing early in the season.

Methods

A tree nursery using gutters to decrease growing costs.

Nurseries grow plants in open fields, on container fields, in tunnels or greenhouses. In open fields, nurseries grow decorative trees, shrubs and herbaceous perennials. On a containerfield nurseries grow small trees, shrubs and herbaceous plants, usually destined for sales in garden centers.These have proper ventilation, sunlight etc. Plants may be grown by seeds. The most common method is by cutting plants/plant cuttings. These can be taken from shoot tips or from roots etc. By these methods plants are grown in nurseries and gardens

Conditioning

With the objective of fitting planting stock more ably to withstand stresses after outplanting, various nursery treatments have been attempted or developed and applied to nursery stock. Buse and Day, for instance, studied the effect of conditioning of white spruce and black spruce transplants on their morphology, physiology, and subsequent performance after outplanting. Root pruning, wrenching, and fertilization with potassium at 375 kg/ha were the treatments applied. Root pruning and wrenching modified stock in the nursery by decreasing height, root collar diameter, shoot:root ratio, and bud size, but did not improve survival or growth after planting. Fertilization reduced root growth in black spruce but not of white spruce.

Frost Hardiness

Seedlings vary in their susceptibility to injury from frost. Damage can be catastrophic if "unhardened" seedlings are exposed to frost. Frost hardiness may be defined as the minimum temperature at which a certain percentage of a random seedling population will survive or will sustain a given level of damage. The term LT_{50} (lethal temperature for 50% of a population) is commonly used. Determination of frost hardiness in Ontario is based on electrolyte leakage from mainstem terminal tips 2 cm to 3 cm long in weekly samplings. The tips are frozen then thawed, immersed in distilled water, the electrical conductivity of which depends on the degree to which cell membranes have been ruptured by freezing releasing electrolyte. A −15 °C frost hardiness level has been used to determine the readiness of container stock to be moved outside from the greenhouse, and −40 °C has been the level determining readiness for frozen storage.

In an earlier technique, potted seedlings were placed in a freezer chest and cooled to some level for some specific duration; a few days after removal, seedlings were assessed for damage using various criteria, including odour, general visual appearance, and examination of cambial tissue.

Stock for fall planting must be properly hardened-off. Conifer seedlings are considered to be hardened off when the terminal buds have formed and the stem and root tissues have ceased growth. Other characteristics that in some species indicate dormancy are color and stiffness of the needles, but these are not apparent in white spruce.

Forest Tree Nurseries

Whether in the forest or in the nursery, seedling growth is fundamentally influenced by soil fertility, but nursery soil fertility is readily amenable to amelioration, much more so than is forest soil.

Nitrogen, phosphorus, and potassium are regularly supplied as fertilizers, and calcium and magnesium are supplied occasionally. Applications of fertilizer nitrogen do not build up in the soil to develop any appreciable storehouse of available nitrogen for future crops. Phosphorus and potassium, however, can be accumulated as a storehouse available for extended periods.

Fertilization permits seedling growth to continue longer through the growing season than unfertilized stock; fertilized white spruce attained twice the height of unfertilized. High fertility in the rooting medium favours shoot growth over root growth, and can produce top-heavy seedlings ill-suited to the rigors of the outplant site. Nutrients in oversupply can reduce growth or the uptake of other nutrients. As well, an excess of nutrient ions can prolong or weaken growth to interfere with the necessary development of dormancy and hardening of tissues in time to withstand winter weather.

Stock Types, Sizes and Lots

Nursery stock size typically follows the normal curve when lifted for planting stock. The runts at the lower end of the scale are usually culled to an arbitrary limit, but, especially among bareroot stock, the range in size is commonly considerable. Dobbs and McMinn examined how the performance of 2+0 bareroot white spruce related to differences in initial size of planting stock. The stock was regraded into large, medium, and small fractions according to fresh weight. The small fraction (20% of the original stock) had barely one-quarter of the dry matter mass of the large fraction at the time of outplanting. Ten years later, in the blade-scarified site, seedlings of the large fraction had almost 50% greater stem volume than had seedlings of the small fraction. Without site preparation, large stock were more than twice the size of small stock after 10 years.

Nursery of apricot seedlings.

Similar results were obtained with regraded 2+1 transplants sampled to determine root growth capacity. The large stock had higher RGC as well as greater mass than the small stock fraction.

The value of large size at the time of planting is especially apparent when outplants face strong competition from other vegetation, although high initial mass does not guarantee success. That the growth potential of planting stock depends on much more than size seems clear from the indifferent success of the transplanting of small 2+0 seedlings for use as 2+1 "reclaim" transplants. The size of bareroot white spruce seedlings and transplants also had a major influence on field performance.

The field performance among various stock types in Ontario plantations was examined by Paterson and Hutchison: the white spruce stock types were 2+0, 1.5+0.5, 1.5+1.5, and 3+0. The nursery stock was grown at Midhurst Forest Tree Nursery, and carefully handled through lifting on 3 lift dates, packing, and hot-planting into cultivated weed-free loam. After 7 years, overall survival was 97%, with no significant differences in survival among stock types. The 1.5+1.5 stock with a mean height of 234 cm was significantly taller by 18% to 25% than the other stock types. The 1.5+1.5 stock also had significantly greater dbh than the other stock types by 30-43%. The best stock type was 57 cm taller and 1 cm greater in dbh than the poorest. Lifting date had no significant effect on growth or survival.

High elevation sites in British Columbia's southern mountains are characterized by a short growing season, low air and soil temperatures, severe winters, and deep snow. The survival and growth of Engelmann spruce and subalpine fir outplanted in 3 silvicultural trials on such sites in gaps of various sizes were compared by Lajzerowicz

Survival after 5 or 6 years decreased with smaller gaps. Height and diameter also decreased with decreasing size of gap; mean heights were 50 cm to 78 cm after 6 years, in line with height expectations for Engelmann spruce in a high-elevation planting study in southeastern British Columbia. In the larger gaps (≥1.0 ha), height increment by year 6 was ranging from 10 cm to 20 cm. Lajzerrowicz Concluded that plantings of conifers in clearcuts at high elevations in the southern mountains of British Columbia are likely to be successful, even close to timberline; and group selection silvicultural systems based on gaps 0.1 ha or larger are also likely to succeed. Gaps smaller than 0.1 ha do not provide suitable conditions for obtaining adequate survival or for growth of outplanted conifers.

Planting Stock

Planting stock, "seedlings, transplants, cuttings, and occasionally wildings, for use in planting out," is nursery stock that has been made ready for outplanting. The amount of seed used in white spruce seedling production and direct seeding varies with method.

A working definition of planting stock quality was accepted at the 1979 IUFRO Workshop on Techniques for Evaluating Planting Stock Quality in New Zealand: "The quality of planting stock is the degree to which that stock realizes the objectives of management (to the end of the rotation or achievement of specified sought benefits) at minimum cost. Quality is fitness for purpose." Clear expression of objectives is therefore prerequisite to any determination of planting stock quality. Not only does performance have to be determined, but performance has to be rated against the objectives of management. Planting stock is produced in order to give effect to the forest policy of the organization.

A distinction needs to be made between "planting stock quality" and "planting stock performance potential" (PSPP). The actual performance of any given batch of outplanted planting stock is determined only in part by the kind and condition, i.e., the intrinsic PSPP, of the planting stock.

The PSPP is impossible to estimate reliably by eye because outward appearance, especially of stock withdrawn from refrigerated storage, can deceive even experienced foresters, who would be offended if their ability were questioned to recognize good planting stock when they saw it. Prior to Wakeley's demonstration of the importance of the physiological state of planting stock in determining the ability of the stock to perform after outplanting, and to a considerable extent even afterwards, morphological appearance has generally served as the basis for estimating the quality of planting stock. Gradually, however, a realization developed that more was involved. Tucker for instance, after assessing 10-year survival data from several experimental white spruce plantations in Manitoba noted that "Perhaps the most important point revealed here is that certain lots of transplants performed better than others", even though all transplants were handled and planted with care. The intuitive "stock that looks good must be good" is a persuasive, but potentially dangerous maxim. That greatest of teachers, Bitter Experience, has often enough demonstrated the fallibility of such assessment, even though the corollary "stock that looks bad must be bad" is likely to be well founded. The physiological qualities of planting stock are hidden from the eye and must be revealed by testing. The potential for survival and growth of a batch of planting stock may be estimated from various features, morphological and physiological, of the stock or a sample thereof.

The size and shape and general appearance of a seedling can nevertheless give useful indications of PSPP. In low-stress outplanting situations, and with a minimized handling and lifting-planting cycle, a system based on specification for nursery stock and minimum morphological standards for acceptable seedlings works tolerably well. In certain circumstances, benefits often accrue from the use of large planting stock of highly ranked morphological grades. Length of leading shoot, diameter of stem, volume of root system, shoot:root ratios, and height:diameter ratios have been correlated with performance under specific site and planting conditions. However, the concept that larger is better negates the underlying complexities. Schmidt-Vogt, for instance, found that whereas mortality among large outplants is greater than among small in the year of planting, mortality in subsequent growing seasons is higher among small outplants than among large. Much of the literature on comparative seedling performance is clouded by uncertainty as to whether the stocks being compared share the same physiological condition; differences invalidate such comparisons.

Height and root-collar diameter are generally accepted as the most useful morphological criteria and are often the only ones used in specifying standards.Quantification of root system morphology is difficult but can be done, e.g. by using the photometric rhizometer to determine intercept area, or volume by displacement or gravimetric methods.

Planting stock is always subject to a variety of conditions that are never optimal in toto. The effect of sub-optimal conditions is to induce stress in the plants. The nursery manager aims, and is normally able to avoid stresses greater than moderate, i.e., restricting stresses to levels that can be tolerated by the plants without incurring serious damage. The adoption of nursery regimes to equip planting stock with characteristics conferring increased ability to withstand outplanting stresses, by managing stress levels in the nursery to "condition" planting stock to increase tolerance to various post-planting environmental stresses, has become widespread, particularly with containerized stock.

Outplanted stock that is unable to tolerate high temperatures occurring at soil surfaces will fail to establish on many forest sites, even in the far north. Factors affecting heat tolerance were investigated by Colombo et al.; the production and roles of heat shock proteins (HSPs) are important in this regard. HSPs, present constitutively in black spruce and many other, perhaps most, higher plants are important both for normal cell functioning and in a stress response mechanism following exposure to high, non-lethal temperature. In black spruce at least, there is an association between HSPs and increased levels of heat tolerance. Investigation of the diurnal variability in heat tolerance of roots and shoots in black spruce seedlings 14 to 16 weeks old found in all 4 trials that shoot heat tolerance was significantly greater in the afternoon than in the morning. The trend in root heat tolerance was similar to that found in the shoots; root systems exposed to 47 °C for 15 minutes in the afternoon averaged 75 new roots after a 2-week growth period, whereas only 28 new roots developed in root systems similarly exposed in the morning. HSP_{73} was detected in black spruce nuclear, mitochondrial, microsomal, and soluble protein fractions, while HSP_{72} was observed only in the soluble protein fraction. Seedlings exhibited constitutive synthesis of HSP_{73} at 26 °C in all except the nuclear membrane fraction in the morning; HSP levels at 26 °C in the afternoon were higher than in the morning in the mitochondrial and microsomal protein factions. Heat shock affected the abundance of HSPs depending on protein fraction and time of day. Without heat shock, nuclear membrane-bound HSP_{73} was absent from plants in the morning and only weakly present in the afternoon, and heat shock increased the abundance of nuclear membrane. Heat shock also affected the abundance of HSP_{73} in the afternoon, and caused HSP_{73} to appear in the morning. In the mitochondrial and microsomal protein fractions, an afternoon heat shock reduced HSP_{73}, whereas a morning heat shock increased HSP_{73} in the mitochondrial but decreased it in the microsomal fraction. Heat shock increased soluble $HSP_{72/73}$ levels in both the morning and afternoon. In all instances, shoot and root heat tolerances were significantly greater in the afternoon than in the morning.

Planting stock continues to respire during storage even if frozen. Temperature is the major factor controlling the rate, and care must be taken to avoid overheating. Navratil found that closed containers in cold storage averaged internal temperatures 1.5 °C to 2.0 °C above the nominal storage temperature. Depletion of reserves can be estimated from the decrease in dry weight. Cold-stored 3+0 white spruce nursery stock in northern Ontario had lost 9% to 16% of dry weight after 40 days of storage. Carbohydrates can also be determined directly.

The propensity of a root system to develop new roots or extend existing roots cannot be determined by eye, yet it is the factor that makes or breaks the outcome of an outplanting operation. The post-planting development of roots or root systems of coniferous planting stock is determined by many factors, some physiological, some environmental. Unsatisfactory rates of post-planting survival unrelated to the morphology of the stock, led to attempts to test the physiological condition of planting stock, particularly to quantify the propensity to produce new root growth. New root growth can be assumed to be necessary for successful establishment of stock after planting, but although the thesis that RGC is positively related to field performance would seem to be reasonable, supporting evidence has been meager.

The physiological condition of seedlings is reflected by changes in root activity. This is helpful in determining the readiness of stock for lifting and storing and also for outplanting after storage. Navratil reported a virtually perfect ($R^2 = 0.99$) linear relationship in the frequency of 3+0 white spruce white root tips longer than 10 mm with time in the fall at Pine Ridge Forest Nursery, Alberta, decreasing during a 3-week period to zero on October 13 in 1982. Root regenerating research with white spruce in Canada followed similar lines to that of Stone's pioneering work in California.

Simpson and Ritchie debated the proposition that root growth potential of planting stock predicts field performance; their conclusion was that root growth potential, as a surrogate for seedling vigor, can predict field performance, but only under such situations as site conditions permit. Survival after planting is only partly a function of an outplant's ability to initiate roots in test conditions; root growth capacity is not the sole predictor of plantation performance.

Some major problems militate against greater use of RGC in forestry, including: unstandardized techniques; unstandardized quantification; uncertain correlation between quantified RGC and field performance; variability within given, nominally identical, kinds of planting stock; and the irrelevance of RGC test values determined on a sub-sample of a parent population that subsequently, before it is planted, undergoes any substantive physiological or physical change. In its present form, RGC testing is silviculturally useful chiefly as a means of detecting planting stock that, while visually unimpaired, is moribund.

Seedling moisture content can be increased or decreased in storage, depending on various factors including especially the type of container and the kind and amount of moisture-retaining material present. When seedlings exceed 20 bars PMS in storage, survival after outplanting becomes problematical. The Relative Moisture Content of stock lifted during dry conditions can be increased gradually when stored in appropriate conditions. White spruce (3+0) packed in Kraft bags in northern Ontario increased RMC by 20% to 36% within 40 days.

Bareroot 1.5+1.5 white spruce were taken from cold storage and planted early in May on a clear-felled boreal forest site in northeastern Ontario. Similar plants were potted and kept in a greenhouse. In outplanted trees, maximum stomatal conductances (g) were initially low (<0.01 cm/s), and initial base xylem pressure potentials (PSIb) were -2.0 MPa. During the growing season, g increased to about 0.20 cm/s and PSIb to -1.0 MPa. Minimum xylem pressure potential (PSIm) was initially -2.5 MPa, increasing to -2.0 MPa on day 40, and about -1.6 MPa by day 110. During the first half of the growing season, PSIm was below turgor loss point. The osmotic potential at turgor loss point decreased after planting to -2.3 MPa 28 days later. In the greenhouse, minimum values of PSIT were -2.5 MPa (in the first day after planting. the maximum bulk modulus of elasticity was greater in white spruce than in similarly treated jack pine and showed greater seasonal changes. Relative water content (RWC) at turgor loss was 80-87%. Available turgor (TA), defined as the integral of turgor over the range of RWC between PSIb and xylem pressure potential at the turgor loss point) was 4.0% for white spruce at the beginning of the season compared with 7.9% for jack pine, but for the rest of the season TA for jack pine was only 2%, to 3% that of white spruce. Diurnal turgor (Td), the integral of turgor over the range of RWC between PSIb and PSIm, as a percentage of

TA was higher in field-planted white spruce than jack pine until the end of the season.

The stomata of both white and black spruce were more sensitive to atmospheric evaporative demands and plant moisture stress during the first growing season after outplanting on 2 boreal sites in northern Ontario than were jack pine stomata, physiological differences that favoured growth and establishment being more in jack pine than in the spruces.

With black spruce and jack pine, but not with white spruce, Grossnickle and Blake's findings warrant mention in relation to the bareroot-containerized debate. During the first growing season after outplanting, containerized seedlings of both species had greater needle conductance than bareroot seedlings over a range of absolute humidity deficits. Needle conductance of containerized seedlings of both species remained high during periods of high absolute humidity deficits and increasing plant moisture stress. Bareroot outplants of both species had a greater early season resistance to water-flow through the soil–plant–atmosphere continuum (SPAC) than had containerized outplants. Resistance to water flow through the SPAC decreased in bareroot stock of both species as the season progressed, and was comparable to containerized seedlings 9 to 14 weeks after planting. Bareroot black spruce had greater new-root development than containerized stock throughout the growing season.

The greater efficiency of water use in newly transplanted 3-year-old white spruce seedlings under low levels of absolute humidity difference in water-stressed plants immediately after planting helps explain the commonly observed favourable response of young outplants to the nursing effect of a partial canopy. Silvicultural treatments promoting higher humidity levels at the planting microsite should improve white spruce seedling photosynthesis immediately after planting.

Stock Types (Seedling Nomenclature)

Planting stock is grown under many diverse nursery culture regimes, in facilities ranging from sophisticated computerized greenhouses to open compounds. Types of stock include bareroot seedlings and transplants, and various kinds of containerized stock. For simplicity, both container-grown and bareroot stock are generally referred to as seedlings, and transplants are nursery stock that have been lifted and transplanted into another nursery bed, usually at wider spacing. The size and physiological character of stock vary with the length of growing period and with growing conditions. Until the technology of raising containerized nursery stock bourgeoned in the second half of the twentieth- century, bareroot planting stock classified by its age in years was the norm.

Classification by Age

The number of years spent in the nursery seedbed by any particular lot of planting stock is indicated by the 1st of a series of numbers. The 2nd number indicates the years subsequently spent in the transplant line, and a zero is shown if indeed there has been no transplanting. A 3rd number, if any, would indicate the years subsequently spent after a second lifting and transplanting. The numbers are sometimes separated by dashes, but separation by plus sign is more logical inasmuch as the sum of the individual numbers gives the age of the planting stock. Thus 2+0 is 2-year-old seedling planting stock that has not been transplanted, and Candy's white spruce 2+2+3 stock had spent 2 years in the seedbed, 2 years in transplant lines, and another 3 years in transplant lines after a second transplanting. Variations have included such self-explanatory combinations, such as 1½+1½, etc.

The class of planting stock to use on a particular site is generally selected on the basis of historical record of survival, growth, and total cost of surviving trees. In the Lake States, Kittredge concluded that good stock of 2+1 white spruce was the smallest size likely to succeed and was better than larger and more expensive stock when judged by final cost of surviving trees.

Classification by Seedling Description Code

Because age alone is an inadequate descriptor of planting stock, various codes have been developed to describe such components of stock characteristics as height, stem diameter, and shoot: root ratio. A description code may include an indication of the intended planting season.

Physiological Characteristics

Neither age classification nor seedling description code indicate the physiological condition of planting stock, though rigid adherence to a given cultural regime together with observation of performance over a number of years of planting can produce stock suitable for performing on a "same again" basis.

Classification by System

Planting stock is raised under a variety of systems, but these have devolved generally into 2 main groupings: bareroot and containerized. Manuals specifically for the production of bareroot and containerized nursery stock are valuable resources for the nursery manager. As well, a lot of good information about nursery stock specific to regional jurisdictions is well presented by Cleary for Oregon, Lavender for British Columbia, and Wagner and Colombo for Ontario.

Thinning

Thinning is a term used in agricultural sciences to mean the removal of some plants, or parts of plants, to make room for the growth of others but does not involve the cutting of the whole tree. Selective removal of parts of a plant such as branches, buds, or roots is typically known as pruning.

Apples thinning.

In forestry, thinning is the selective removal of trees, primarily undertaken to improve the growth rate or health of the remaining trees. Overcrowded trees are under competitive stress from their neighbors. Thinning may be done to increase the resistance of the stand to environmental stress such as drought, insect infestation, extreme temperature, or wildfire.

Thinning in the Umatilla National Forest.

This may be done to make the stand more profitable in an upcoming final felling or to achieve ecological goals such as increasing biodiversity or accelerating the development of desired structural attributes such as large diameter trees with long tree crowns.

Early thinning, say after 20 years rather than late thinning, say after 50 years has different effects on the trees thinned. An early thinning would encourage trees to get wider crowns, increase diameter and be more stable against threats like snow breakage or windthrow. Too much too early would lead to a site overgrowing with shrubs and prevent an understory regenerating effectively. Doing it too late would mean that trees would grow tall and slender and though some increment would come to the trees, they may be less marketable as thinness reduces value. Too early and epicormic shooting is a risk, and this can lead to tree branchiness and knots in the resultant timber harvest, again reducing value.

Traditionally thinning has been done to create a desired balance between individual tree attributes (such as tree diameter) and per area attributes such as volume. It has been, and often still is, applied with the desire to create uniform stands. Thinning treatments are often described in terms of number of trees per area to remain or average spacing between trees. The intent is to create and manage uniform stands. It is also more needed when too many trees were initially planted or survived from the seedling phase. Planting less and thinning less saves money in commercial forestry, so a good balance is required and is dependent on site qualities as well as planting régimes.

There is no sure result from thinning and no standard timing or other aspect. Individual site conditions and species responses or species mixes on sites may vary considerably and thus responses to thinnings likewise. Norway spruce on a suitable site have responded well in terms of growth increment, to late thinnings whereas other species have not for example slash pine.

Due to such variability it is better to talk about a thinning régime rather than one particular method of thinning taking place in a stand.

Methods

- Thinning from below: This low thinning can be split into 4 Grades: A Grade is a very light thinning, that removes all overtopped trees Kraft crown class 4 and 5. B Grade is a very light thinning that removes overtopped trees and intermediates which are Kraft Crown class 4,5 and some 3s, C Grade and D Grade are a moderate and heavy thinning respectively removing anything that will not lead to high quality tree crown growth.
- Thinning from above: This crown thinning removes all of the trees that impact crown wise on other trees. Smaller trees, Kraft crown class 5 for example would just be ignored and left in the stand and not cut. It is also called the French method.
- Diameter-limit thinning this selection thinning is related to the Borggreve silvicultural system. Large trees over a certain diameter are removed for smaller trees in good condition to grow. These in turn after some growth are thinned. Eventually the stand is exhausted after several cycles and the remnant is clear cut and replanted. It is used with shade tolerant species like Western hemlock.
- Geometrical thinning: Mechanical thinning like this is used in plantations where one row is taken out and the next left no matter the tree condition. A different grid might be used for natural stands sometimes called strip thinning.
- Free thinning: This is done as crop tree release and it means thinning around a wanted tree, over a stand this may mean in some areas no thinning and in others a lot of thinning. This heterogeneity represents where the most valuable trees are.

Ecological Thinning

Fresh snow partially covers epixylic Rough-stalked Feather-moss (Brachythecium rutabulum), growing on a dead hybrid black poplar (Populus x canadensis), thinned in 2008 to create a waterside path. The last stage of the moss lifecycle is shown, where the sporophytes are visible before dispersion of their spores:

- The calyptra is still attached to the sporangium.
- The tops of the gametophytes.
- Can be discerned as well.

Inset shows the surrounding, black poplars growing on sandy loam on the bank of a kolk, with the detail area marked.

Ecological thinning is a silvicultural technique used in forest management that involves cutting trees to improve functions of a forest other than timber production.

Although thinning originated as a man-made forest management tool, aimed at increasing timber yields, the shift from production forests to multifunctional forests brought with it the cutting of trees to manipulate an ecosystem for various reasons, ranging from removing non-native species from a plot to removing poplars growing on a riverside beach aimed at recreational use.

Since the 1970s, leaving the thinned trees on the forest floor has become an increasingly common policy: wood can be decomposed in a more natural fashion, playing an important role in increasing biodiversity by providing habitat to various invertebrates, birds and small mammals. Many fungi (e.g. Calocera viscosa) and mosses are saproxylic or epixylic as well (e.g. Marchantiophyta) – some moss species completing their entire life-cycle on a single log.

Where trees are managed under a commercial regime, competition is reduced by removing adjacent stems that exhibit less favourable timber quality potential. When left in a natural state trees will "self-thin", but this process can be unreliable in some circumstances. Examples of this can be found in the Buxus – Ironbark forests and woodlands of Victoria (Australia) where a large proportion of trees are coppice, resultant from timber cutting in decades gone by.

Ecophysiological Repercussions

Thinning decreases canopy closure and increases the penetration of solar radiation into the canopy. The photosynthetic efficiency of this energy is improved, and needle retention is prolonged, especially in the lower parts of the crown. The root system, crown length, crown diameter, and crown area all increase after thinning. Even if soil evaporation and individual tree transpiration increases after thinning, total evapo-transpiration at stand level tends to decrease; canopy water interception is reduced and throughfall increased, so that tree-water status usually improves after thinning.

Radial annual growth is an integrative index of tree physiological response to environmental variation. Working with Norway spruce, which can be expected to behave in some respects similarly to white spruce, Misson showed that a reduction in stand density alters the classical climate–growth relationship. At individual tree level, thinning could be used to increase tree resistance to drought stress. Nevertheless, this effect is limited when site conditions are limiting. Misson concluded that heavy thinning should be applied when forest decline is expected from drought stress. Furthermore, stands on dry sites should be thinned more heavily because such sites cannot support high-density stands.

Aussenac, also working with Norway spruce, investigated the response to thinning. Thinning altered the temporal evolution of radial growth at all frequencies in radial growth chronologies. Previous studies had shown, not surprisingly, that thinning decreases the canopy closure and encourages solar radiation to penetrate the canopy. Solar energy then becomes more important inside the crown. The improved photosynthetic efficiency of this energy influences the retention time of needles, especially in the lower part of the crown. Furthermore, the length, diameter, and area, of the crown, and the size of the root system all increase after thinning.

These modifications at the crown level influence photosynthate production positively as long as water supply is not limiting. Even if soil evaporation and individual tree transpiration are more important after thinning, total evapo-transpiration at stand level tends to decrease. Furthermore, with the reduced importance of leaf area index (LAI), thinning reduces canopy water interception and increases throughfall. This explains why the soil moisture content increases in a thinned stand. Thus, during the vegetation period, tree-water status is usually better in a thinned stand than in a high-density stand.

For semi-tolerant species such as Norway spruce and white spruce, solar energy is less limiting than for intolerant species. This can account for the exponential relationship found by Misson between thinning intensity and mean radial growth. Only when thinning was relatively severe did important radial growth variation occur. Furthermore, the ecophysiological advantage of an improved water supply could be counterbalanced by limiting site conditions.

For Norway spruce in the Belgian Ardennes, Misson recommended that stand basal area should not exceed 26 m^2/ha on a dry site, or 29 m^2/ha on a moist site if the objective is to maintain high long-term radial growth.

Misson found that adaptations to thinning cover a continuous range of different time scales, with long-, medium-, and short-term growth variations being controlled by very different internal factors influenced by the environment. Misson ascribed what they called long- and medium-term radial growth variation of individual trees mainly to structural adaptation, such as enlargement of the crown or the root system. Short-term radial growth variations, on the other hand, were considered to be due mainly to physiological acclimatisation, with response to factors such as stomatal conductance regulation, variation of photosynthetic capacity, and respiration. Nevertheless, it is clear that the adaptations are linked, and that radial growth is the integrative response of this mutual inter-relationship.

Pruning

Pruning: Dense growth after shearing.

Pruning is a horticultural and silvicultural practice involving the selective removal of certain parts of a plant, such as branches, buds, or roots. Reasons to prune plants include deadwood removal, shaping (by controlling or redirecting growth), improving or sustaining health, reducing risk from falling branches, preparing nursery specimens for transplanting, and both harvesting and increasing the yield or quality of flowers and fruits.

The practice entails targeted removal of diseased, damaged, dead, non-productive, structurally unsound, or otherwise unwanted tissue from crop and landscape plants. In general, the smaller the branch that is cut, the easier it is for a woody plant to compartmentalize the wound and thus limit the potential for pathogen intrusion and decay. It is therefore preferable to make any necessary formative structural pruning cuts to young plants, rather than removing large, poorly placed branches from mature plants.

Specialized pruning practices may be applied to certain plants, such as roses, fruit trees, and grapevines. It is important when pruning that the tree's limbs are kept intact, as this is what helps the tree stay upright. Different pruning techniques may be deployed on herbaceous plants than those used on perennial woody plants. Hedges, by design, are usually (but not exclusively) maintained by hedge trimming, rather than by pruning.

In nature, meteorological conditions such as wind, ice and snow, and salinity can cause plants to self-prune. This natural shedding is called abscission.

Pruning Landscape and Amenity Trees

Types of Branch Union

For arboricultural purposes the unions of tree branches (i.e. where they join together) are placed in one of three types: collared, collarless or codominant. Regardless of the overall type of pruning being carried out, each type of union is cut in a particular way so that the branch has less chance of regrowth from the cut area and best chance of sealing over and compartmentalising decay. This is often referred to by arborists as "target cutting".

Pruning when there's a branch collar: The swollen area where the branch joins the trunk is known as the collar.

Pruning when there's a collarless union: The area where there is no swollen area where the branch joins the trunk is called a collarless union.

Pruning when it's a codominant stem.

Deadwooding

Branches die off for a number of reasons including light deficiency, pest and disease damage, and root structure damage. A dead branch will at some point decay back to the parent stem and fall off. This is normally a slow process but can be quickened by high winds or extreme temperature. The main reason deadwooding is performed is safety. Situations that usually demand removal of deadwood is trees that overhang public roads, houses, public areas and gardens. Trees located in wooded areas are usually assessed as lower risk but assessments consider the number of visitors. Usually, trees adjacent to footpaths and access roads are considered for deadwood removal.

Another reason for deadwooding is amenity value, i.e. a tree with a large amount of deadwood throughout the crown looks more aesthetically pleasing with the deadwood removed. The physical practice of deadwooding can be carried out most of the year though not when the tree is coming into leaf. The deadwooding process speeds up the tree's natural abscission process. It also reduces unwanted weight and wind resistance and can help overall balance.

Crown and Canopy Thinning

Crown and canopy thinning increases light and reduces wind resistance by selective removal of branches throughout the canopy of the tree.

Crown Canopy Lifting

Crown lifting involves the removal of the lower branches to a given height. The height is achieved by the removal of whole branches or removing the parts of branches which extend below the desired height. The branches are normally not lifted to more than one third of the tree's total height.

Crown lifting is done for access; these being pedestrian, vehicle or space for buildings and street furniture. Lifting the crown will allow traffic and pedestrians to pass underneath safely. This pruning technique is usually used in the urban environment as it is for public safety and aesthetics rather than tree form and timber value.

Crown lifting introduces light to the lower part of the trunk; this, in some species can encourage epicormic growth from dormant buds. To reduce this sometimes smaller branches are left on the lower part of the trunk. Excessive removal of the lower branches can displace the canopy weight; this will make the tree top heavy, therefore adding stress to the tree. When a branch is removed from the trunk, it creates a large wound. This wound is susceptible to disease and decay, and could lead to reduced trunk stability. Therefore, much time and consideration must be taken when choosing the height the crown is to be lifted to.

This would be an inappropriate operation if the tree species' form was of a shrubby nature. This would therefore remove most of the foliage and would also largely unbalance the tree. This procedure should not be carried out if the tree is in decline, poor health or dead, dying or dangerous (DDD) as the operation will remove some of the photosynthetic area the tree uses. This will increase the decline rate of the tree and could lead to death.

If the tree is of great importance to an area or town, (i.e. veteran or ancient) then an alternative solution to crown lifting would be to move the target or object so it is not in range. For example, diverting a footpath around a tree's drip line so the crown lift is not needed. Another solution would be to prop up or cable-brace the low hanging branch. This is a non-invasive solution which in some situations can work out more economically and environmentally friendly.

- Directional or Formative Pruning: Removal of appropriate branches to make the tree structurally sound while shaping it.
- Vista Pruning: Selectively pruning a window of view in a tree.

Crown Reduction

Reducing the height and or spread of a tree by selectively cutting back to smaller branches and in fruit trees for increasing of light interception and enhancing fruit quality.

Pollarding

A regular form of pruning where certain deciduous species are pruned back to pollard heads every year in the dormant period. This practice is usually commenced on juvenile trees so they can adapt to the harshness of the practice.

Types

Arborists, orchardists, and gardeners use various garden tools and tree cutting tools designed for the purpose, such as hand pruners, loppers, or chainsaws. Additionally in forestry bush saws are commonly used and these are often attached to poles that reach up to 5-6m, this is a more efficient way of pruning than with ladders. These bush saws on polls have also been motorized as chainsaws which is even more efficient. Older technology used Billhooks, Kaiser blades and pruning knives. Although still used in some coppicing they are not used so much in commercial forestry due to the difficulty of cutting flush with the stem. Flush cuts lead to good wood, non-flush or bark damaging cuts (which are more likely with a swung blade than a sawed one) put the tree at risk of entry cords from forest pathogens.

An arborist pruning a tree near the Statue of Liberty.

Regardless of the various names used for types of pruning, there are only two basic cuts: One cuts back to an intermediate point, called heading back cut, and the other cuts back to some point of origin, called thinning out cut.

Removing a portion of a growing stem down to a set of desirable buds or side-branching stems. This is commonly performed in well trained plants for a variety of reasons, for example to stimulate growth of flowers, fruit or branches, as a preventative measure to wind and snow damage on long stems and branches, and finally to encourage growth of the stems in a desirable direction. Also commonly known as heading-back.

- Thinning: A more drastic form of pruning, a thinning out cut is the removal of an entire shoot, limb, or branch at its point of origin. This is usually employed to revitalize a plant by removing over-mature, weak, problematic, and excessive growths. When performed correctly, thinning encourages the formation of new growth that will more readily bear fruit and flowers. This is a common technique in pruning roses and for amplifying and "opening-up" the branching of neglected trees, or for renewing shrubs with multiple branches.
- Topping: Topping is a very severe form of pruning which involves removing all branches and growths down to a few large branches or to the trunk of the tree. When performed correctly it is used on very young trees, and can be used to begin training younger trees for pollarding or for trellising to form an espalier.

- Raising removes the lower branches from a tree in order to provide clearance for buildings, vehicles, pedestrians, and vistas.
- Reduction reduces the size of a tree, often for clearance for utility lines. Reducing the height or spread of a tree is best accomplished by pruning back the leaders and branch terminals to lateral branches that are large enough to assume the terminal roles (at least one-third the diameter of the cut stem). Compared to topping, reduction helps maintain the form and structural integrity of the tree.

In orchards, fruit trees are often lopped to encourage regrowth and to maintain a smaller tree for ease of picking fruit. The pruning regime in orchards is more planned and the productivity of each tree is an important factor.

Deadheading is the act of removing spent flowers or flowerheads for aesthetics, to prolong bloom for up to several weeks or promote rebloom, or to prevent seeding.

Time Period

In general, pruning dead wood and small branches can be done at any time of year. Depending on the species, many temperate plants can be pruned either during dormancy in winter, or, for species where winter frost can harm a recently pruned plant, after flowering is completed. In the temperate areas of the northern hemisphere autumn pruning should be avoided, as the spores of disease and decay fungi are abundant at this time of year.

Some woody plants tend to bleed profusely from cuts, such as mesquite and maple. Some callus over slowly, such as magnolia. In this case, they are better pruned in during active growth when they can more readily heal. Woody plants that flower early in the season, on spurs that form on wood that has matured the year before, such as apples, should be pruned right after flowering as later pruning will sacrifice flowers the following season. Forsythia, azaleas and lilacs all fall into this category.

Clearcutting

Clearcutting, clearfelling or clearcut logging is a forestry/logging practice in which most or all trees in an area are uniformly cut down. Along with shelterwood and seed tree harvests, it is used by foresters to create certain types of forest ecosystems and to promote select species that require an abundance of sunlight or grow in large, even-age stands. Logging companies and forest-worker unions in some countries support the practice for scientific, safety and economic reasons, while detractors consider it a form of deforestation that destroys natural habitats and contributes to climate change.

After a century of clearcutting, this forest, near the source of the Lewis and Clark River in Clatsop County, Oregon, is a patchwork. In each patch, most of the trees are the same age.

Clearcutting is the most common and economically profitable method of logging. However, it also creates detrimental side effects, such as the loss of topsoil, the costs of which are intensely debated by economic, environmental and other interests. In addition to the purpose of harvesting wood, clearcutting is used to create land for farming. The human demand for wood and arable land through unsustainable logging regimes like clearcutting has led to the loss of over half of the world's rainforests.

While deforestation of both temperate and tropical rainforests through clearcutting has received considerable media attention in recent years, the other large forests of the world, such as the taiga, also known as boreal forests, are also under threat of rapid development. In Russia, North America and Scandinavia, creating protected areas and granting long-term leases to tend and regenerate trees—thus maximizing future harvests—are among the means used to limit the harmful effects of clearcutting. Long-term studies of clearcut forests, such as studies of the Pasoh Rainforest in Malaysia, are also important in providing insights into the preservation of forest resources worldwide.

Types

Many variations of clearcutting exist; the most common professional practices are:

- Standard (uniform) clearcut: Removal of every stem (whether commercially viable or not), so no canopy remains.
- Patch clearcut: Removal of all the stems in a limited, predetermined area (patch).
- Strip clearcut: Removal of all the stems in a row (strip) usually placed perpendicular to the prevailing winds in order to minimize the possibility of windthrow.
- Clearcutting-with-reserves: Removal of the majority of standing stems save a few reserved for other purposes (for example as snags for wildlife habitat), (often confused with the seed tree method).
- Slash-and-burn: The permanent conversion of tropical and subtropicals forests for agricultural purposes. This is most prevalent in tropical and subtropical forests in overpopulated regions in developing and least developed countries. Slash-and-burn entails the removal of all stems in a particular area. This can be a form of deforestation, when the land is converted to other uses. However,

some indigenous forest peoples, for example the 19th century Forest Finns rotate over the land and it does return to forest and this would be sustainable. Slash and burn techniques are typically used by civilians in search of land for living and agricultural purposes. The forest is first clear cut, and the remaining material is burned. One of the driving forces behind this process is a result of overpopulation and subsequent sprawl. These methods also occur as a result of commercial farming. The lumber is sold for profit, and the land, cleared of all remaining brush and suitable for agricultural development, is sold to farmers.

- Selection cutting: which can be done for timber harvesting or for ecological reasons. when so done it is often called ecoforestry.

Clearcutting contrasts with selective cutting, such as high grading, in which only commercially valuable trees are harvested, leaving all others. This practice can reduce the genetic viability of the forest over time, resulting in poorer or less vigorous offspring in the stand. Clearcutting also differs from a coppicing system, by allowing revegetation by seedlings. Additionally, destructive forms of forest management are commonly referred to as 'clearcutting'.

Clearcutting Regeneration, Harvesting or System

Clearcutting can be differentiated into:

- Clearcutting: Clean felling by complete exploitation and removal of all the trees in one operation, a harvesting method.
- Clearcutting method: Method for regenerating an even-aged community by removing all the mature trees.
- Clearcutting system: A silvicultural system incorporating the clearcutting method to remove (clear) the mature community over a considerable area at one time.

Clearcut logging in the Blue Ridge Mountains.

Clearcutting in Southern Finland.

Clearcutting near Eugene, Oregon.

Confusion between these different uses of the term is common. Furthermore, as indicated above many variations mean technically correct usage may not be descriptive enough to know what is meant on that particular occasion.

Effects on the Environment

Environmental groups criticize clear-cutting as destructive to water, soil, wildlife, and atmosphere, and recommend the use of sustainable alternatives. Clear-cutting has a very big impact on the water cycle. Trees hold water and topsoil. Clear-cutting in forests removes the trees which would otherwise have been transpiring large volumes of water and also physically damages the grasses, mosses, lichens, and ferns populating the understorey. All this bio-mass normally retains water during rainfall. Removal or damage of the biota reduces the local capacity to retain water, which can exacerbate flooding and lead to increased leaching of nutrients from the soil. The maximum nutrient loss occurs around year two, and returns to pre-clearcutting levels by year four.

Clear-cutting also prevents trees from shading riverbanks, which raises the temperature of riverbanks and rivers, contributing to the extinction of some fish and amphibian species. Because the trees no longer hold down the soil, riverbanks increasingly erode as sediment into the water, creating excess nutrients which exacerbate the changes in the river and create problems miles away, in the sea. All of the extra sediment and nutrients that leach into the streams cause the acidity of the stream to increase, which can kill marine life if the increase is great enough. The nutrient content of the soil was found to return to five percent of pre-clearcutting levels after 64 years, which demonstrates how clearcutting affects the environment for many years.

Clearcutting can destroy an area's ecological integrity in a number of ways, including: the destruction of buffer zones which reduce the severity of flooding by absorbing and holding water; the immediate removal of forest canopy, which destroys the habitat for many rainforest-dependent insects and bacteria; the removal of forest carbon sinks, leading to global warming through the increased human-induced and natural carbon dioxide build-up in the atmosphere; the elimination of fish and wildlife species due to soil erosion and habitat loss; the removal of underground worms, fungi and bacteria that condition soil and protect plants growing in it from disease; the loss of small-scale economic opportunities, such as fruit-picking, sap extraction, and rubber tapping; and the destruction of aesthetic values and recreational opportunities.

Upper town residential area of Britannia Beach in British Columbia, Canada in September 1919. The community was built in a clearcut area adjacent to the Britannia Mines operations; stumps and stray logs are visible throughout the site and community.

Negative Impacts

Clearcutting can have major negative impacts, both for humans and local flora and fauna. A study from the University of Oregon found that in certain zones, areas that were clear cut had nearly three times the amount of erosion due to slides. When the roads required by the clearcutting were factored in, the increase in slide activity appeared to be about 5 times greater compared to nearby forested areas. The roads built for clearcutting interrupt normal surface drainage because the roads are not as permeable as the normal ground cover. The roads also change subsurface water movement due to the redistribution of soil and rock. Clearcutting may lead to increased stream flow during storms, loss of habitat and species diversity, opportunities for invasive and weedy species, and negative impacts on scenery, as well as a decrease in property values; diminished recreation, hunting, and fishing opportunities. Clearcutting decreases the occurrence of natural disturbances like forest fires and natural uprooting. Over time, this can deplete the local seed bank. An example of what clearcutting did in Ontario before 1900 can be found in Edmund Zavitz.

In temperate and boreal climates, clearcutting can have an effect on the depth of snow, which is usually greater in a clearcut area than in the forest, due to a lack of interception and evapotranspiration. This results in less soil frost, which in combination with higher levels of direct sunlight results in snowmelt occurring earlier in the spring and earlier peak runoff.

The world's rain forests could completely vanish in a hundred years at the current rate of deforestation. Between June 2000 and June 2008 more than 150 000 square kilometers of rain forest were cleared in the Brazilian Amazon. Huge areas of forest have already been lost. For example, only eight to fourteen percent of the Atlantic Forest in South America now remains. While deforestation rates have slowed since 2004, forest loss is expected to continue for the foreseeable future. Farmers slash and burn large parcels of forest every year to create grazing and crop lands, but the forest's nutrient-poor soil often renders the land ill-suited for agriculture, and within a year or two, the farmers move on.

Positive Perspectives

Clearcutting can be practiced to encourage the growth and proliferation of tree species that require high light intensity. Generally, a harvest area wider than double the height of the adjacent trees will no longer be subject to the moderating influence of the woodland on the microclimate. The width of the harvest area can thus determine which species will come to dominate. Those with high tolerance to extremes in temperature, soil moisture, and resistance to browsing may be established, in particular secondary successional pioneer species.

Clearcutting can be used by foresters as a method of mimicking a natural disturbance and increasing primary successional species, such as poplar (aspen), willow and black cherry in North America. Clearcutting has also proved to be effective in creating animal habitat and browsing areas, which otherwise would not exist without natural stand-replacing disturbances such as wildfires, large scale windthrow, or avalanches.

Clearcuts are used to help regenerate species that cannot compete in mature forests. A number of them are aspen, jack pine and, in areas with poor soils, oaks—are important species for both game and nongame wildlife species. Clearcutting can also lead to increased vascular-plant diversity in the area. This is most pronounced after a couple years of clearcutting and in herb-rich forests where scarification took place.

No significant changes in water temperature were observed when patch clearcutting was done 100 feet away from a river. This suggests that patch clearcutting is a possible solution to concerns about changes in water temperature due to clearcutting.

More recently, forest managers have found that clearcutting oak stands helps regenerate oak forests in areas of poor soil. The tree canopies in oak forests often shade out the ground, making it impossible for newly sprouted oaks to grow. When the mature trees are removed, the saplings stand a chance of recruiting into the forest.

Effects on Wildlife

Clearcutting's main destruction is towards habitats, where it makes the habitats more vulnerable in the future to damage by insects, diseases, acid rain, and wind. Removal of all trees from an area destroys the physical habitats of many species in wildlife. Also clearcutting can contribute to problems for ecosystems that depend on forests, like the streams and rivers which run through them.

In Canada, the black-tailed deer population is at further risk after clearcutting. The deer are a food source for wolves and cougars, as well as First Nations and other hunters.

While deer may not be at risk in cities and rural countryside, where they can be seen running through neighbourhoods and feeding on farms, in higher altitude areas they require forest shelter.

Variable Retention

Variable retention is a relatively new silvicultural system that retains forest structural elements for at least one rotation in order to preserve environmental values associated with structurally complex forests.

Some examples for environmental values are forest cover connectivity, soil stabilization, microclimate under retained trees, habitats associated with live or dead trees and species diversity due to habitat preservation, and wildlife corridor preservation. Variable retention also aids in emulation of natural disturbances by leaving behind some residual structure from previous stand which is typical for stand replacing disturbances Traditional silviculture systems such as clearcut, patch cut, shelterwood, etc. are focused on maximizing timber production and future regeneration of the trees. Variable retention on the other hand is focused on what is retained.

Value of Variable Retention

Variable retention is gradually becoming a popularly promoted tool for helping resolve the dilemma between demands for wood and demands to maintain habitat and ecology biodiversity and structural diversity in managed forests. What is at issue is how much a forest operation can log without adversely interfering with other ecological processes within the forest.

Variable retention minimizes the impact of logging operation by leaving biological legacies such as coarse woody debris (nurse logs and snags). Either few trees or many trees can be retained under the variable retention system, and trees can be retained in patches (aggregated retention) or left uniformly throughout a stand (dispersed retention); hence the name "variable retention." It is a technique for retaining trees as key structural elements of a harvested stand for at least until next harvest rotation in an effort to maintain species, habitat diversity and forest-related processes. A 2014 meta-analysis of retention forestry research found "support for wider use of retention forestry since it moderates negative harvesting impacts on biodiversity."

There are four key mechanisms through which variable retention is presumed to maintain biodiversity:

- By providing a constant supply of structural features that are at high risk to being lost due to modern forestry practices and that are known to be important to habitat availability, such as large trees, very young trees, snags, and coarse woody debris.

- By providing adequate refuge for sensitive species that will colonize the surrounding managed forest environment as it develops suitable conditions.
- By establishing habitat patches, Patch dynamics, that can serve as stepping stones for the dispersal of newly produced offspring, seeds, and spores.
- By increasing the structural diversity of managed stands.

Variable Retention - Harvesting Methods

The method of variable retention is a more deliberate process than clearcutting. In some cases helicopters take the logs out. The use of the helicopters avoid the need to build logging roads inside the forest and hence avoid the disturbance of the forest floor. Yet, helicopter logging is less fuel efficient than truck use. Some logging operations that use variable retention try to emulate forest disturbances. In those cases, deadwood in snags and logs is left behind. Such dead or dying trees provide a growth medium for fungi and epiphytes, as well as food and shelter for insects and the birds. Loggers also leave younger trees that will continue to grow.

Variable retention harvesting retains more than 15% of the original stand in both rolling and permanent pockets of untouched trees and refugia patches. Retained forest structure is composed of live and dead trees that enhance structural diversity, and provide a hospitable habitat for variety of living organisms that prevail in the unlogged forest. This silviculture regime provides post-harvest ecological structure while creating sufficient opportunity to plant and naturally regenerate valuable tree species for timber, as well as restore historical coniferous and deciduous tree dominance to the forestland. Some timber companies have restricted the use of variable retention silviculture to only poorly stocked stands of rare but valuable tree species.

Controversy

Ecologists regard the variable retention method of harvesting as being far less detrimental to the forest ecosystem than clearcutting. However, opponents claim that variable retention is much more laborious, tedious, time-consuming and expensive than clearcutting. A review of studies in British Columbia and in similar forest types indicates that while both aggregated and dispersed retention can contribute to biodiversity conservation there may be some negative consequences for timber production such as wind damage to retained trees and reduced growth rates of tree regeneration compared to clearcuts, concluding in favour of an adaptive management approach for balancing competing objectives of production of wood and biodiversity conservation.

Forestry organizations, such as the Forest Stewardship Council (FSC), approve of variable retention and are now declaring timber obtained through such a harvesting method as certified wood for green building. Many forest operations in the Canadian province of British Columbia, including MacMillan Bloedel, use the variable retention method to harvest timber in the temperate rain forests along the Pacific coast.

Many logging companies have not maintained a total commitment to phasing out clearcutting and embracing variable retention harvesting, and even MacMillan Bloedel had occasionally fallen short of its commitment to phase out clearcutting. Although world markets are beginning to ask for certified wood obtained entirely through variable retention harvesting, it is not yet clear whether the general population of consumers will be willing to pay higher prices for such wood.

Coppicing

Coppicing is the process of cutting trees down, allowing the stumps to regenerate for a number of years (usually 7 - 25) and then harvesting the resulting stems.

It makes use of the natural regeneration properties of many tree species, including Oak, Hazel, Maple, Sweet Chestnut, Lime and Ash. Cut such trees down and they will regenerate from the cut stump, producing many new shoots, rather than a single main stem. Regrowth can be exceedingly rapid, with new shoots growing as much as 5cm a day. Oak stems can exceed 2m growth in one season, while Sallow may grow to almost 4m high in the first summer1. The word 'coppice' is derived from the French 'couper' which means 'to cut'. Coppice trees and their produce are known as 'underwood'.

The cut tree stump is known as a stool and the shoots, depending on their harvested size, as rods, poles or logs. The shoots are harvested on a rotational cycle. This means that they are left to grow for a certain number of years and are then cut, whereupon the whole process starts again.

Coppiced woods are either cut according to demand or will more usually be divided into a number of compartments. One of the compartments, or coupes, will be cut each year within the rotational cycle.

The length of the rotational cycle will depend on the tree species, as well as the projected use of the product. A typical rotation might be on a cycle of 7 - 8 years. Many traditional uses require 7 - 15 year old material, although some modern commercial use may take larger material, up to 30 years old.

Coppicing has been practiced in British woodlands for centuries. As a result of the rotational cutting sequence, at any one time there would be coppice at various different stages of regeneration within the woodland. In this way, wood was produced for a variety of uses, in an elegantly sustainable way.

Long rotations were used which would provide really sizeable timber. Coppice products included ships planking, timber for half-timbered Tudor houses, as well as smaller items such as pea and bean sticks, firewood, charcoal, furniture, sheep hurdles, baskets, fencing, hedging sticks, tool handles and brooms.

References

- What-is-silviculture: maineforestry.net, Retrieved 25 June, 2020
- Nyland, Ralph D. (2002). Silviculture: Concepts and Applications. McGraw-Hill series in forest resources (2nd ed.). Boston: McGraw-Hill. p. 20. ISBN 978-0-07-366190-2
- Silvicultural-systems-sp, silviculture-silviculture-textbook-sp-28339: forestry.sfasu.edu, Retrieved 08 January, 2020
- Mitchell, S.J.; Beese, W.J. (2002). "The retention system: Reconciling variable retention with the principles of silvicultural systems". Forestry Chronicle. 78 (3): 397–403. doi:10.5558/tfc78397-3. hdl:10613/2698
- Coppice, woodland-manage: countrysideinfo.co.uk, Retrieved 10 March, 2020

6

Benefits of Sustainable Forestry

Carbon sequestration, biodiversity conservation, conservation of soil, protection of water quality, improvement in air quality, reduction in soil erosion, etc. are some of the benefits of sustainable forestry. This chapter delves into these benefits of sustainable forestry for a thorough understanding of the subject.

Carbon Sequestration

The best way to manage forests to store carbon and to mitigate climate change is hotly debated. Trees absorb carbon dioxide from the atmosphere, and wood can be a substitute for fossil fuels and carbon-intensive materials such as concrete and steel. In the past few decades, the world's forests have absorbed as much as 30% (2 petagrams of carbon per year; Pg C year^{-1}) of annual global anthropogenic CO_2 emissions – about the same amount as the oceans. Two-thirds of forests are managed.

Much has been learned about the carbon cycle in forests, but there are still too many gaps in our knowledge. New observations have called long-accepted theories into question: the finding that unharvested forests, for example, are absorbing more carbon than they release, accounting for half the sink, is contrary to the tenet of ecology, known as Odum's framework, that carbon flows in natural forests should be in equilibrium.

This carbon-sink behaviour of mature forests is attributed to large-scale environmental changes that violate the assumption of the steady conditions underlying Odum's framework: higher atmospheric CO_2 concentrations are accelerating tree growth worldwide and nitrogen emitted by industry, agriculture and fossil-fuel burning is increasingly fertilizing managed forest soils in Europe, China and the eastern United States.

To make good decisions about how to cultivate forests for climate-change mitigation, such as whether it is better to harvest or conserve trees, we must better understand the cause and future behaviour of this in situ carbon sink. Until more is known, we propose that forestry management should prioritize 'win–win' strategies — those that increase both forest stocks and timber harvest, through measures such as protecting trees from animals, or replacing dying or low-productivity forests.

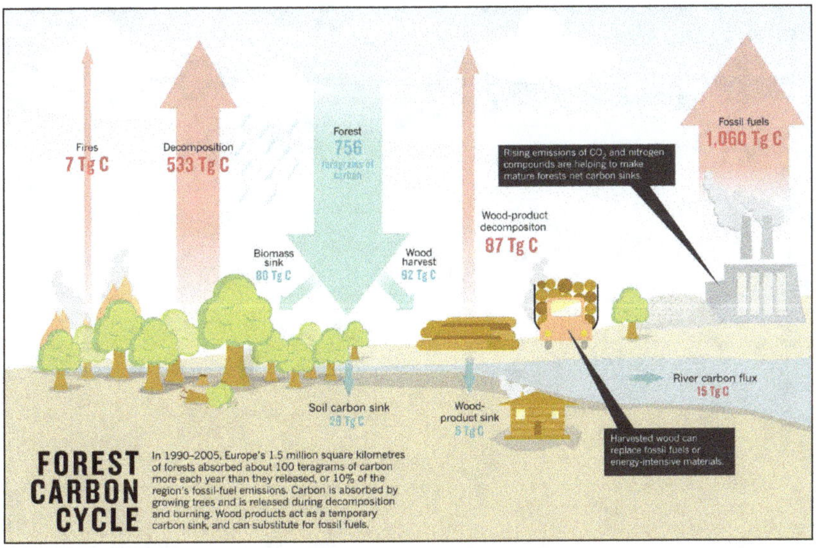

Global Drivers

So far, most discussions about the impact of climate change on forests have focused on the rising rate of locally devastating events such as forest fires, infestations, droughts and storms. The past decade saw 410 million cubic metres of wood felled in four major storms in Europe, a decade's worth of Amazonian carbon sequestration lost in severe droughts in 2005 and 2010, a record heat wave with forest fires affecting 23,000 square kilometres in Russia in 2010, and the bark beetle pandemic that has affected 130,000 square kilometres and killed 435 million cubic metres of trees in British Columbia, Canada, since 2004.

Regionally, these events have had large socio-economic and ecological impacts. The windstorms that hit Europe in 1999, for example, destroyed one-third of the region's annual carbon sequestration by forests, and halved the price of timber in France and other affected countries. But such events are insignificant over decades and at the global scale. The Pacific El Niño and La Niña climate oscillations — primarily the tropical forest and peatland fires associated with them — are the only regional disturbances evident in globally averaged atmospheric CO_2 records.

At the same time, atmospheric records and forest inventories show that forests have been taking up ever more CO_2 in the past 50 years. Experimental and modelling studies have established that currently, the global forest sink is driven mainly by changes in atmospheric CO_2 concentration and in nitrogen deposition. The competing explanation, that today's rapid tree growth is a recovery from ancient climate or harvesting losses seems unlikely because it requires that regions around the globe, such as the Amazon and the Congo Basin, were simultaneously affected. Such a coincidence should have left traces of soot and enhanced CO_2 concentrations in air bubbles in glacial ice core samples, and these are not seen.

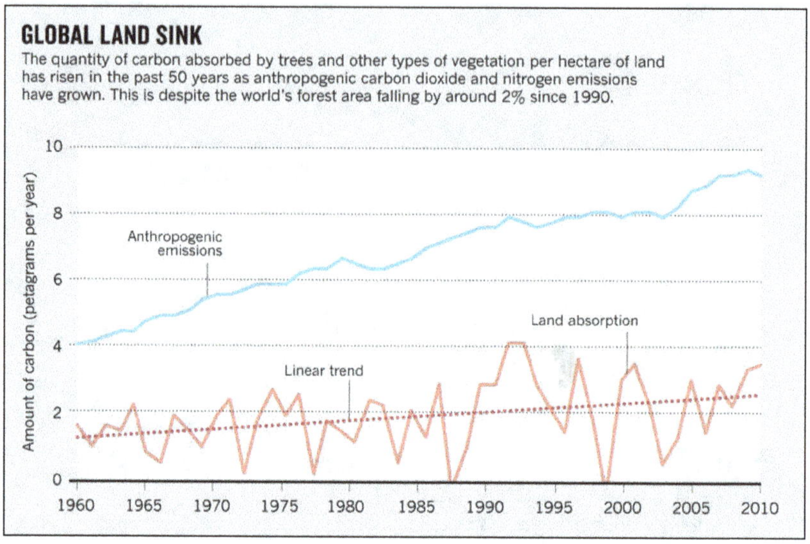

The size of the global forest carbon sink has increased alongside rising CO_2 levels. Industrial and agricultural activities and fossil-fuel burning emit nitrogen compounds that fertilize forests up to hundreds of kilometres away. The rate of carbon storage in temperate forests in Europe and North America is well correlated with their exposure to this nitrogen deposition. Other changing climate factors, such as temperature and rainfall, and changing forest-management strategies, such as leaving trees to grow for longer before cutting them back, seem to be of secondary importance to the global carbon budget of forests, although they may be locally important.

A quantified understanding of how all these drivers shape the forest carbon sink is lacking. And predictions of how they will change during this century remain uncertain. Projections that CO_2 might overtake nitrogen deposition as the dominant cause of the sink in temperate zones, as industrial emissions fall, are controversial. Physiological constraints, such as trees living shorter lives as their growth accelerates and soil fertility falling in mature woodlands, are often missing from models, as is the effect of temperature on tree growth.

As a result, the picture is confused. Different models disagree on whether the forest carbon balance in 2100 will be positive or negative, let alone its magnitude. Even where models make the same assumptions, such as no change in current emissions and no forest management, they have wildly different predictions. Models assuming that rising temperatures and CO_2 concentrations will increase photosynthesis, which absorbs CO_2 faster than respiration emits CO_2, suggest that the biosphere could absorb up to 10 Pg C $year^{-1}$. This is five times today's terrestrial carbon sink and matches current CO_2 emissions from fossil-fuel burning and deforestation. If respiration dominates, others predict that the biosphere will become a carbon source of 6 Pg C year^{-1}, doubling the current CO_2 emissions and thus requiring emissions reductions far beyond what is being discussed.

Three Ways Out

The future trajectory of the carbon sink influences how forests should be managed for climate-change mitigation. If the world's forests remain net absorbers, conservation would be appealing. Preserved mature forests would absorb almost as much carbon as younger ones. Because decomposing harvest residues and roots add immediately to the CO_2 emissions, and it takes decades for increased use of wood products to compensate, avoiding harvest could generate extra climate benefits, at least in the short run.

Harvesting reduces carbon emissions from decomposition.

Conversely, if mature forests become carbon sources, increased harvesting may be the best mitigation option. Harvesting would reduce losses from decomposition while promoting wood as a fossil-fuel substitute.

While a scientifically robust prediction of the persistence of the forest carbon sink is worked out, three safe strategies should be followed.

First, so that studies can be compared and uncertainties addressed, scientists should state their assumptions more clearly. The Earth system modelling community should define and report a set of internationally agreed performance quality criteria, based on an ability to reproduce trends in large data sets that include interactions relevant to understanding the effects of nitrogen and CO_2 on the forest sink. To provide these data sets, large-scale remote sensing needs to be combined with frequent monitoring of local test sites, such as the permanent plots of national forest inventories, with monitoring networks such as Fluxnet (which coordinates measurements of CO_2 exchanges), and with controlled experiments, such as CO_2-enrichment studies.

The forest-science community should make explicit the assumed behaviour of unmanaged forests that lies behind its assessment of forest mitigation strategies and its life-cycle analysis for wood products. Most of these studies assume that unmanaged forests are carbon neutral, which overestimates the climate benefits of harvest. Others assume that the *in situ* forest sink will be sustained forever, underestimating the climate benefits of harvest.

Second, the most carbon-efficient uses of wood should be encouraged. Harvesting more timber could be, especially if the forest sink starts shrinking, a good climate-change mitigation strategy, but to be effective it must be targeted to uses that will save the most tonnes of CO_2 per cubic metre harvested. In construction, for example, wood can substitute for steel or cement for studs or walls, and can then be recovered, recycled and burnt.

Third, forest-management techniques that increase both the amount of wood produced and the carbon stock retained in the forest should be prioritized. When not in conflict with other forest uses, replacing dying or low-productivity stands, protecting young sprouts from damage after harvest, planting tree mixes that are more resilient, and optimizing fertilizer use and tree growth by adding nitrogen-fixing species in afforestation projects, will contribute to climate-change mitigation no matter how the global carbon sink evolves.

Biodiversity Conservation

The multiple-use forest management model has been flexible, versatile, and useful in accommodating a growing array of forest uses and evolving social values regarding forest conservation and stewardship. The need to protect biological diversity in forest ecosystems presents a difficult challenge to this model, however, due to its importance and to the fact that adequate protection of many threatened or endangered species is not compatible with even moderate levels of resource development. An effective global strategy for biodiversity conservation must consider biodiversity values in temperate forest nations relative to those in tropical and boreal ecosystems, along with the potential of temperate-forest nations to minimize the effects of their own high per-capita demand for wood fibber on the most diverse, globally-significant ecosystems. Developed temperate-forest nations have a dual conservation responsibility to:

- Protect their own globally significant biodiversity hotspots.
- Minimize externalities by more intensive utilization of productive forest areas with relatively low biodiversity values.

Ultimately, the overarching concept of "sustainable forestry" must be inclusive of reserves and intensively managed plantations, as well as forests managed at moderate intensity for a wide variety of goods, services, and values.

Most of the world's biodiversity will continue to exist outside protected areas and there are also managed lands within many protected areas. In the assessment of millennium targets, there is therefore a need for indicators to measure biodiversity and suitability of habitats for biodiversity both across the whole landscape/seascape and in specific managed habitats. The two predominant land uses in many inhabited areas are forestry and agriculture and these are examined. Many national-level criteria and indicator systems already exist that attempt to assess biodiversity in forests and the impacts of forest management, but there is generally less experience in measuring these values in agricultural landscapes.

The seventh meeting of the Conference of the Parties of the Convention on Biological Diversity (CBD), in February 2004, identified a series of trial indicators to be developed and used to report on the previously agreed target for CBD member states: to 'achieve by 2010 a significant reduction of the current rate of biodiversity loss at the global, regional and national levels as a contribution to poverty alleviation and to the benefit of all life on Earth.' Two indicators relating to sustainable use were identified as 'possible indicators for development', aimed for adoption at the next Conference of Parties in Brazil in early 2005, giving an 18-month period to develop realistic and agreed ways for their measurement. The two indicators provisionally identified are:

- Area of forest, agricultural and aquaculture ecosystems under sustainable management.
- Proportion of products derived from sustainable sources.

These two measures reflect efforts to measure and develop sustainable use, that is, a response to biodiversity loss rather than being a direct measure of biodiversity itself, although there is clearly some overlap between these two.

Protected Areas or Sustainable Use

It is sometimes suggested that the existence of scale of protected areas could serve as surrogate indicators for biodiversity conservation status and that; therefore, the survival of biodiversity in forests and farmland is relatively unimportant. We reject this, both because most of the world's biodiversity continues to exist outside protected areas and because many protected areas also contain managed land.

Protected areas are the cornerstones of most national and international conservation strategies, providing refuges for species that cannot survive and ecological processes that cannot be maintained in intensely managed landscapes or seascapes. Today, there are over 100000 designated protected areas listed by the World Database on Protected Areas (WDPA covering over 12% of the Earth's land surface). Most of these were identified and gazetted during the twentieth century, in what is probably the largest conscious land use change in history. However, this also means that almost 90% of the world's land surface still remains outside formal protected areas, themselves subject to varying degrees of biodiversity protection in practice. This would be less important in the context of biodiversity monitoring if the world's biodiversity was mainly concentrated in protected areas, but in fact the majority remains outside. Although a few species, such as the two African rhinoceros species, are now virtually confined to various categories of state-run protected areas or private land (including hunting areas), these are the exceptions rather than the rule. The giant panda (Ailuropoda melanoleuca) provides a good example of the tendency for wild nature to retain a foothold in the wider landscape. The Government of China has had the technical expertise and the political power to establish a series of protected areas that focus specifically on those places with healthy populations of pandas in areas where there is otherwise quite intense competition between people and wildlife. Yet, despite these ideal conditions for representing biodiversity within the protected areas network, surveys show that the majority of giant pandas continue to live outside protected areas. Similarly, in Hawaii the ranges of birds that are most rare exist outside the protected area network. A global analysis of vertebrates found at least 12% of species not represented in any protected area. For more mobile species, this mismatch is often the rule.

Protected areas also only function effectively as tools for conservation if they are well managed and they retain their constituent species and habitats. Unfortunately, many protected areas are currently under threat or are experiencing degradation and loss. Research in 10 forest-rich countries found that local experts only rated 1% of protected areas as fully secure and almost a quarter were already suffering some form of serious degradation. A global survey found widespread threats, particularly in the tropical countries. Surveys in 93 protected areas in 22 tropical countries and in 206 protected areas in temperate and tropical countries found several critical threats, in particular, illegal use related to logging and poaching. While these surveys suggest that the most protected areas continue to protect some biodiversity, no assessment system should assume that protected area status is equivalent to effective biodiversity conservation.

Furthermore, protected areas surrounded entirely by radically altered habitat have limited usefulness in the long term for many species unless the areas are very large. Species trapped in protected area 'islands' risk genetic isolation and gradual decline. For example, in Java, Indonesia, the botanical gardens in Bogor were isolated in 1936 when surrounding forest was destroyed. Despite the garden's forests being maintained intact, bird populations have declined. Between 1932 and 1952, 62 species of birds were recorded in the gardens but, by the 1980s, 20 species had disappeared, four were close to extinction and five more had declined substantially . The concept of protected areas as networks surrounded by buffer zones and interconnected by corridors and stepping stones of semi-natural habitat, suitable at the very least for the passage of species to allow genetic interchange, is now accepted throughout the conservation community and was a major theme of the Fifth World Parks Congress, held in Durban in September 2003. Increasingly, conservation organizations are looking at conservation on the scale of ecoregions and landscapes, which explicitly cover far more than just networks of protected areas.

In addition, protected areas are not necessarily made up entirely of untouched habitat. Protected landscapes and seascapes (IUCN Category V) and extractive reserves (IUCN Category VI) both may contain a considerable proportion of their area devoted to some kind of agricultural or forest management. These constitute a large area: 28.9% of the total protected areas that have been given an IUCN category, covering well over 5 million km^2.

If biodiversity is to be conserved outside protected area networks, in economically productive landscapes, this implies that biodiversity use is sustainable in the overall landscape and in addition that management is compatible with the survival of some or all of the biodiversity originally present. Article 2 of the CBD defines sustainable use as: 'The use of components of biological diversity in a way and at a rate that does not lead to the long-term decline of biological diversity, thereby maintaining its potential to meet the needs and aspirations of present and future generations'.

Sustainability must therefore be ensured from the environmental, social and economic viewpoints. Environmental sustainability implies that exploited populations should not be reduced to densities where they can no longer fulfil their ecological role—as pollinators, seed dispersers or predators, as part of the food chain, etc.—and conversely should be harvested when their densities rise above a level where their ecological role becomes distorted. Disrupting these functions potentially results in a gradual but profound shift in ecosystem functions and processes. Economic sustainability means that biodiversity should not be reduced to a point where it ceases to fulfil its economic roles in ensuring sustained livelihoods of dependent people. Lastly, social sustainability relates to the wide-ranging and diverse benefits of biodiversity for local communities that play an important role in addressing many social demands. Human-related disturbances (from global to local, from industries to communities) and policy decisions should be managed in a way that sustain or increase these benefits.

From a practical perspective, a sustainable use is one use that can be maintained in the long term. The statement that sustainable use is a form of conservation has therefore some merit and local interest in the resource is an important incentive for its maintenance. It should be clear also that all uses, consumptive or non-consumptive, will impact on the ecology in some way and will translate into more or less dramatic effects on the local environment, depending on the nature of this activity.

It is essential to note that the optimal population density is likely to be different in each case outlined above (i.e. environmental, economic or social sustainability), and also to consider the time factor. As nobody can be sure that a particular use will be sustained indefinitely, there can only ever be a probability of a use being sustainable, based on current knowledge and commitments or on a qualification of the estimated sustainability by its expected duration and the necessary conditions.

Accurate measurement of biodiversity status therefore needs to embrace the whole landscape mosaic, and to include consideration of biodiversity values in managed lands both inside and outside protected areas. Ideally, this should include the full range of different managed habitats but the two most significant terrestrial management systems: forests and agricultural land, the latter covering both cropland and grazing areas.

Critical Challenges Facing Biodiversity Measurement at a Large Scale

While this broader scale of analysis is essential, it will certainly not be easy. Attempts at measuring progress in global biodiversity targets across a whole landscape, in a way that allows us to detect trends over a relatively short six-year timescale, will have to solve a number of serious methodological problems. At least three critical challenges remain to be addressed.

Variable Data Quality

Unfortunately, there is often an inverse relation between the quality of data about biodiversity within a region or country and the conservation value of biodiversity itself; in general, the places with the best records are also those with the most impoverished wildlife. Thus, countries such as the UK and The Netherlands have unusually detailed information about wild nature, but an almost entirely cultural landscape where top predators are long extinct, little or no fully natural habitat remains and a largely unmeasured proportion of biodiversity has already disappeared. Countries such as those in the Congo Basin or the Amazon still contain huge areas of largely unmodified habitat but only the most superficial information even about the largest species and a significant proportion of smaller plants and animals have never been described. The risk from the perspective of the 2010 target is that the only reliable trend data will come from the least important areas.

Such a level of data poverty makes it almost impossible to set a realistic baseline against which to measure progress. Attempts to set up a monitoring system for a conservation project covering seven provinces of central Vietnam faced just such a problem: population levels of wild species remain virtually unknown, even for megafauna such as the tiger. The differences are stark. In the UK, most groups are now mapped at a 10km² level, with a very high degree of coverage for mammals, birds, butterflies and flowering plants (with some counties now mapping on a 2km² level). On the other hand, when WWF International assembled 130 experts for an ecoregional assessment of the Congo Basin, they failed to identify the presence or absence of elephants over an area the size of France and Germany combined.

Indicators often do not Include Information on Biodiversity It self

Many 'biodiversity indicators' actually give little information about overall biodiversity levels or status. Most monitoring, particularly in data poor countries, focuses on large species, usually mammals and birds or on land cover types. While these may give general information about the 'existence' of habitat, they often say comparatively little about its 'quality' (including its ability to support other species). Forest elephants in both Africa and Asia, for instance, can continue to live in quite degraded forests as long as sufficient food plants remain and the same is true for many of the great apes such as gorilla (Gorilla beringei) and orang utan (Pongo pygmaeus). Forests can appear intact or only slightly degraded using remote sensing tools when fauna has been virtually destroyed or while undergoing cryptic deforestation.

Mistakes in the choice of indicator species can distort our understanding of overall biodiversity, either exaggerating or understating the magnitude of the impacts of management and use. For example, using a relatively adaptable species as an indicator can disguise changes to more sensitive species and thus serve to delay necessary management responses. In the middle decades of the twentieth century, the status of reindeer (Rangifer tarandus) and moose (Alces alces) were used as a major indicator of forest condition in managed forests in northern Scandinavia. Both these choices were justifiable: the reindeer has enormous cultural and economic importance to the Sami people and the moose is an important source of food to many families in Arctic Lapland. Yet these large mammals can thrive in secondary forest where many other species will decline or disappear. Continued loss of old growth forest, and in key forest components such as standing and lying dead timber, took place while reindeer and moose populations were maintained, resulting in threats to many species associated with natural forests so that today there are for instance several hundred saproxylic (dead wood living) species on the Red List for Sweden.

In the absence of detailed information about wild species, those seeking practical indicators are often forced to rely on various forms of surrogates, such as habitat condition, umbrella or keystone species or particular management interventions as is already implicit in the suggested indicators from the CBD. These have limitations which should be noted. Habitat condition tells us quite a lot about the likelihood that certain assemblages of species will be present but offers little direct evidence of their presence, and specific pressures such as poaching or air pollution can reduce or eliminate some elements of biodiversity even within a habitat that is apparently pristine. Lack of large mammals in some protected areas in the Congo Basin is a case in point: surveys of vegetation show a healthy forest but some keystone species are absent. Measuring progress by management intervention takes measurement even further away from quantifiable reality by assuming that particular interventions invariably aid biodiversity. In many cases the data are not yet strong enough to support these claims unequivocally. Changes in population densities of umbrella or keystone species can be recorded but we have no real basis to assess what percentage decline or increase is an acceptable level of change.

In practice, until now, the choice of indicators has often been influenced mainly by the particular interests of the specialists involved or the availability of data, rather than being an attempt to provide an overall assessment. Exceptions are some work on identifying useful assemblages and drawing up hierarchies of indicators and identifying indicators for larger areas such as has been attempted for Alaska. Yet a broader understanding of biodiversity implies selecting a suite of indicators that will give as broad a picture as possible of the state of ecology and wild nature.

The need to Choose Indicators that Resonate with the Public and Decision-makers

Although indicators of biodiversity condition should clearly be driven by good science, they also have an important political aspect, which needs to be acknowledged and addressed if they are to be successful in influencing management. Biodiversity indicators are not only important in providing scientists with a means of measuring the state of nature, but they also have to resonate with the policy-makers who respond to the information that they contain, to politicians, industry representatives, land managers, non-governmental organizations and perhaps most important of all, with the wider public (to whom governments are responsible). Indicators in a politically driven process such as the CBD, therefore also have a role in telling a wider story about biodiversity and in setting and driving responses to the pressures that they illuminate. Indicators not only have to be technically coherent and defensible, but to provide information in a way that many people outside the fields of the biological sciences will be able to understand and to sympathize with. To be of political value, an indicator or suite of indicators may need to embody some kind of 'vision' for how the problems it is illuminating could be addressed. This implies a level of negotiation and trade-offs may be needed in reaching a final indicator list.

Agreement on specific indicators within an international agreement implies a certain level of commitment to it being applied and therefore to the message that they contain. As such, good biodiversity indicators in this context need to have more attributes than would be needed from a strictly scientific point of view, and in an ideal world would:

- Be easy to identify and locate in the field without heavy reliance on specialists and often encountered when the management or uses are at an acceptable level.
- Provide reliability but also a broad range of information within a single indicator.
- Give good value for the resources invested (i.e. avoiding data that is more complex or expensive to collect than necessary).
- Be usable by many people and if possible not relying on a small number of specialists to identify or collect.
- Resonate with public and policy makers and telling a clear and understandable story about what is happening within an ecosystem.

This also implies that all indicators will need accompanying explanation, with their limitations clearly articulated. Once indicators leave the confines of the scientific world and are used in wider policy-making and by many more stakeholders, they will also be shaped and sometimes constrained as a result of these wider interests and varying applications.

Indicators can 'foster better management. However, there are confusions and tensions to reconcile between general and local applications, between the ideal and the pragmatic, and between the scientific and the democratic. To overcome this requires a sober appraisal of what can realistically be achieved in each location and how this can best be promoted. Good judgement remains the foundation of competent management. Data can inform this judgement, but an over-reliance on data collection and top-down bureaucratic interventions can add to problems rather than solving them'.

Current Status and Application of Sustainable use Indicators for Terrestrial Systems

There is currently no single source of information that will tell us everything we want to know about sustainable use. To date, considerably more effort has been made to develop global indicators of sustainable forest management than those relating to agriculture and forest management is therefore considered.

Indicators of Sustainable Forest Management

The 1992 Earth Summit in Rio de Janeiro first set international targets for forest conservation through its Forest Principles. Although much derided at the time for being too weak, in retrospect the Principles played an important role because they set a precedent of international targets for forest management. In the years since, other more quantifiable targets have been developed by institutions such as the International Tropical Timber Organisation (ITTO). More are in the process of development by the CBD and the United Nations Forum on Forests, and there are a handful of influential targets advocated by non-governmental organizations. Having agreed to the principle of measurable targets, governments have had to make some effort to find ways of recording these criteria and indicators of good forest management assumed much higher political importance than in the past.

At a global level, a range of criteria and indicators have been developed for tropical forests by the ITTO and for forest biodiversity by the CBD. ITTO indicators are often quite detailed, covering specific management approaches (plantations, natural forests) and there are guidelines relating to biodiversity. However, use of any of the ITTO guidelines has been voluntary and highly erratic and there is no obligation to collect data; although this may change because ITTO is trying to require a reporting system based on indicators in its member countries.

Issues of forest management, including specifically biodiversity, were addressed by the temperate and boreal forest component of the Forest Resource Assessment 2000, organized jointly by the FAO and the UN Economic Commission for Europe. The Temperate and Boreal Forest Resource Assessment 2000 addressed data collection through the use of national correspondents and included considerable details about biodiversity, including information on protected areas, naturalness, species under threat, invasive species and area and type of regeneration. The survey was the first attempt to collect data on area of natural forest as compared with total forest cover. While incomplete and of variable accuracy, TBFRA2000 represents the best attempt to date to identify biodiversity information for the entire forest estate. However, the tropical section of the global report carries far less information about biodiversity, in part because of the way that data are assembled in the tropics, but also, to a large extent, because of genuine difficulties in finding the necessary information. The mismatch between temperate and tropical systems is continued through many other attempts to gather data about forests.

In a parallel process, six regional criteria and indicator processes for sustainable forest management have been developed, drawing on the Forest Principles from the Earth Summit and focusing in greater detail on particular issues relevant to different parts of the world. The first two, the Ministerial Conference on the Protection of Forests in Europe (MCPFE) and the Montreal Process, were initiatives of groups of governments; later developments were coordinated by the Food and Agricultural Organization of the United Nations (FAO) and in one case by the African Timber Organization and the ITTO. Some of these continue to be influential while others have since lost funding and appear to have been abandoned. By far the strongest systems are the Montreal Process and the MCPFE, while most of the tropical criteria and indicator initiatives have either been only weakly applied or not at all, reinforcing the split in data quality between temperate and tropical regions. The work at international and regional level has been complemented by the development of national systems in many temperate countries, in part to implement regional C&I processes.

The latest report from the MCPFE requests that governments report on many issues relating directly to the biodiversity of managed forests, including, for instance, level of threat to forest-dwelling species, regeneration, protection and even the amount of deadwood. While many countries admit to having poor information to respond to some of these questions, efforts are currently underway to improve monitoring schemes and to build a comprehensive European data set on forest condition. Unfortunately, this level of information is not mirrored in any way in most tropical countries.

Lastly, there has been a wide range of non-governmental initiatives. Many are based around stand-level assessments for the purpose of assessing performance of forest management to achieve certification under the auspices of bodies such as the Forest Stewardship Council, the Programme for Endorsement of Forest Certification Schemes (formerly the Pan-European Certification Scheme) or the International Organization for Standardization. These are interesting in that they are established with the specific aim of improving management for social and environmental reasons, including forest biodiversity, and they function through third-party inspection and verification against agreed principles and standards, embarked upon on a voluntary basis by forest managers. This means that there are regular site visits to certified forests and at least some data collection. Certification is also unusual in that it can be a joint effort between forest owners (either state or private) and forest products companies, creating a rare and direct link between industry and biodiversity monitoring.

However, the extent to which various certification schemes focus on forest biodiversity values differs markedly (and also differs between national standards even within a particular umbrella scheme such as the Forest Stewardship Council (FSC) or the Programme for the Endorsement of Forest Certification schemes (PEFC). Indeed, one of the problems in using data from certification schemes in biodiversity monitoring is that there has been a protracted and sometimes bitter dispute about the worth of different schemes. All have attracted criticism from at least some environmental NGOs for not being stringent enough and although some institutions, such as the United Nations Economic Commission for Europe (UNECE), tend to count them all as one in forestry statistics, this would certainly not be supported by all stakeholders. Certification under all schemes still covers less than 5% of the world's forests and less than 10% of certifications are in the tropics. Furthermore, to date, the biodiversity values of such schemes are still largely 'inferred;amp;rsquo; from the changes in management needed rather than 'measured' through long-term studies in changes in forest biodiversity following certification and some NGOs remain sceptical that sustainable forest management offers a good option for conserving forest biodiversity.

There are also some NGO attempts to measure forest quality criteria at a regional level, such as the WWF European forest scorecards and a system of assessment of forest quality developed by WWF, IUCN, the World Conservation Union and the École Polytechnique Fédérale de Lausanne. These all include methodologies for the collection of data on biodiversity, but again have only been applied on a limited scale or in certain parts of the world. More recently, WWF and the World Bank have started to collaborate on a simple scorecard for measuring progress in sustainable forest management in forest stands, building on an existing scorecard used to track progress in protected areas. When complete this will include a rapid process of collecting information on forest management based on around 30 multiple-choice questions.

There are marked similarities between the regional systems, to the extent that there was an unsuccessful attempt to combine them into one system, associated with an intergovernmental meeting in Helsinki in 1996. There have also been a number of attempts to evaluate and compare the various schemes, most notably by the Center for International Forestry Research and on a more theoretical basis by the Dutch-based Tropenbos Foundation. CIFOR has also developed a methodology for users to select a portfolio of indicators suitable to their particular situation and has tested this in many parts of the world.

Indicators of Sustainable Agriculture

Agriculture has generally received less attention than forests with respect to measuring biodiversity value or conservation success, especially at the international level. Many indicators are still partially developed, have never been put into operation or are apparently abandoned. Two exceptions to this are the Organisation for Economic Co-operation and Development (OECD) and the European Union, where a number of initiatives, by the European Commission, EUROSTAT, the European Environment Agency and various EU research programmes, are currently attempting to develop agricultural sustainability indicators, including some for on-farm biodiversity, with the aim of implementation in due course.

The OECD has had a substantial programme of work on agri-environment indicators in recent years, including a number of workshops, working documents and published reports. The aim has been a manageable number of indicators covering the full range of sustainability, including biodiversity, which has proved one of the more technically challenging and politically sensitive. Within the EU, the emphasis has also been on a broad suite of indicators, a small proportion of which are founded on biodiversity. The European Environment Agency is one of the organizations taking forward this work and has also initiated studies exploring the scope for developing indicators of 'high natural value' agriculture in Europe.

Compared with forestry, there is generally less agreement about how the links between biodiversity and agriculture might be measured, with much of the emphasis (where it occurs at all) put towards measuring detrimental impacts of agriculture on surrounding habitats (for instance through soil erosion or pollution run-off) rather than looking at biodiversity within agricultural systems. This is an important shortcoming because, while the most intensive farming systems support little biodiversity, as shown through serious declines in species previously associated with farmland, many traditional agricultural systems, rangelands and various forms of ecologically based agriculture can support a proportion of wild species. Residual natural and semi-natural vegetation and structures on farmland, if not overly affected by inappropriate management, can provide important reservoirs of biodiversity and corridors and stepping-stones between natural habitat including protected areas. Low-intensity farming systems can mimic some of the attributes of more natural systems and can be associated with greater species diversity than other land uses competing with them.

Some indicator systems are now trying to address these wider issues, often based around life-cycle approaches to sustainable agriculture, but there are currently few global or regional data sets available comparable to those associated with sustainable forest management. Certain species, e.g. farmland birds in Europe, have become an indicator for progress informally while a broader international consensus is being sought. There are also data on levels of some key aspects of agricultural management, such as the intensity of nitrogen or agrochemical use or stocking density in different regions and countries.

One exception to the general rule about lack of application and data is the system of certified organic agriculture, operated in many countries and generally meeting principles agreed upon at an international level. The certified management has the advantage of regular inspections and data collection as an intrinsic part of its requirements. Some forms of organic agriculture have also, unlike certified forests, been proven to support more biodiversity than surrounding farms, even if these are managed extensively. Thus, as an indicator, it could act as a fairly strong surrogate for biodiversity conservation. However, the global proportion of land under organic agriculture remains proportionately very small, albeit expanding. The total area under organic management is more than 24 million hectares (ha) worldwide. In addition, the area of certified 'wild harvested plants' is at least a further 10.7 million ha, according to various certification bodies.

Developments in criteria and indicators for sustainable forest management have led to many potential indicators and to a reasonable assemblage of data for many of the temperate and boreal regions, but far less for the tropics. The situation is generally less well-developed for agriculture and, although there is now an array of indicators, again particularly in the richer temperate countries, collection of data is far less advanced.

Possible Framework for Monitoring with Initial Indicators

Given the timescale and likely budget, global indicators need to be based as much as possible on existing information. However, consensus on indicators will be hampered by the major discrepancies in data quality among different regions of the world. Developing a system that every country or even that most countries can meet in the short term would mean setting aside much of the relatively good information available in those countries where assessments have been undertaken for some time. On the other hand, basing data requirements entirely on these northern-based datasets would effectively prevent the participation of most of the countries with the richest biodiversity. The need is thus for a system that makes the best use of available data (and encourages data collectors) without excluding countries with less monitoring infrastructure. In some countries, fairly reliable data based on measurement, will already be available, in others there will be some data but also a measure of value judgements, while in a third group, information may have to draw largely on expert opinion. All these can provide useful information, but they become potentially misleading if they are not distinguished; this is a problem with some official data sources at present, where there is no indication of how far the data draws from systematic surveys or is little better than guesswork.

It is therefore proposed that data for the sustainable use indicators be distinguished according to quality of information that they use, by distinguishing between the accuracy of measurement of particular indicators. This can be done by dividing them into a number of categories: for instance quantitative, semi-quantitative or qualitative; or perhaps some measure of accuracy such as frequency of surveys or use of primary data. (This approach might well be appropriate for other indicators within the CBD targets.)

Each participating country would therefore report on the same indicators but would list the data used in the appropriate column, adding notes if further explanation were required, so that the likely accuracy of this data is made reasonably transparent. This would allow virtually all countries to report on specific indicators but would not set an impossibly ambitious task over the next six years for those nations where data collection has not even begun. Governments could in these cases also work with the CBD and other partners to improve the quality of data (a target which might be attractive to some donor agencies), thus moving to a 'higher' column as time goes on. For data users, it would provide a far more realistic picture and also give an indication of the strength of particular data sources. While governments would be responsible for assigning the level of data reliability and accuracy themselves, independent scientists, NGOs and civil society would, in this case, also play an important role in verifying such judgements.

Suggested Indicators

Indicators measuring sustainable use from a biodiversity perspective often take one of two different approaches: either drawing on information on the area under a sustainable use or on information about the status of biodiversity in these areas. The latter, being an outcome-based indicator, is ultimately more informative but far more difficult to measure. It also provides little direct information about progress in sustainable management (because, for example, biodiversity might show positive trends either if management was improved on an area or it ceased altogether, and the area abandoned to return to a semi-natural state). In the following proposals, we suggest a mixture of both approaches.

The CBD has already proposed some draft indicators for measuring sustainable use in forests and agriculture. The following suggestions draw on these, outline how they might be implemented in practice and make some additional suggestions for indicators that might be added to the initial list in order to complete a more rounded picture of the biodiversity implications of different management approaches. For both forestry and agriculture, a series of indicators could help to both capture and then refine information about sustainable use, which might fall into four general categories as outlined below:

- Proportion of land under sustainable management of specified kinds.
- Amount or value of produce from land under sustainable management.
- Area of natural or high quality semi-natural land within landscapes under sustainable management.
- Key indicator species (assuming that the various caveats of the indicator species approach are addressed.

The first two relate directly to management, but both assume that the sustainability of broad categories of management can be agreed and measured and, furthermore, that this will have net benefits for biodiversity, or will at least not be detrimental. The third goes slightly deeper by looking at land within these use categories that remains in a condition 'likely' to be good for biodiversity, and the last proposes an outcome-related biodiversity indicator for these areas. Data collected for other indicators might also provide information likely to be useful for judging the reality of sustainable use, particularly with respect to monitoring wild species.

Indicators of Sustainable Forest Use

Using this general template, the key indicator should be area of forest land under sustainable management; this is an obvious choice and has also already been identified by the CBD. Adding produce from sustainably managed forests would help to refine the information further by also showing the proportion of traded goods that come from these improved management systems. However, the potential problems with applying these and the limited information that they convey suggest that the overall picture might be further expanded by information on natural forest cover and if possible on key species that indicate good management.

For each of these, agreement is needed on the form that the indicator takes, and on a series of definitional and methodological requirements, along with new survey work.

Table: Development requirements for indicators of sustainable forest use.

Form of indicator	Terms requiring definition	Methodological requirements	Survey requirements	Possible partners	Potential problems
Indicator 1: Area under sustainable forest management					
Either an indicator with a series of stages or a scoring system	'Sustainable management'	Agreement on measuring levels of sustainability	Global surveys	ITTO, World Bank, CIFOR, certification bodies, wood product companies	Agreement on what constitutes sustainable
Indicator 2: Products from sustainably managed forests					
Volume or value of products	'Sustainably managed'	Accounting methods	Existing surveys amalgamated into global figure	Certification bodies, major retailers, FAO	Agreement on what constitutes sustainable
Indicator 3: Area of natural forests					

Proportion of existing forest in natural state	'Natural'	Rapid survey methods (ideally satellite images) in tropics	Data available for most temperate forests, still needed in tropics, Southern Arc and north Asia	FAO, UNECE, ITTO, NASA, World Resources Institute	Identifying natural forests; agreement on what is 'natural' in some areas
Indicator 4: Indicator species					
Status of key species reliant on sustainably managed forests	Identification of indicator species or surrogates	Survey methods exist	Major survey task unless data already collected for other indicators	National biodiversity surveys, Red List, NGOs, universities, research institutes	Costs of identifying and surveying relevant species may be prohibitive

Sustainable forest management could be measured either by completing a series of steps (for instance, existence of good governance, regulatory framework, control of illegal logging and certification of good management) or through some scoring system that uses these as fields in an assessment system. Examples from existing systems, such as the World Bank/WWF scorecard, that is currently under development, and a 'governance pyramid' developed by the International Institute for Environment and Development for use at a national level to indicate quality of forest governance, could help to develop national assessment systems using either of these approaches. The ease with which sustainable trade data could be separated from other statistics for wood products would depend on the definition of sustainable that is adopted. Figures for trade in various forms of certified wood products are already available but if a less stringent definition were agreed upon it would be correspondingly more difficult to separate out the proportion of sustainable products in the market.

Many methods for defining 'natural' forests already exist and data, albeit of varying quality, are available for North America, Europe, the CIS, Japan, Australia and New Zealand through the UNECE Temperate and Boreal Forest Resource Assessment. These increasingly focus on current status rather than ecological history, i.e. assuming that 'naturalness' is an attribute that can be both lost and gained rather than a condition that can only be lost once. Collecting additional data from other countries would present a major challenge, even if satellite data were available (it is not certain how easy it is to distinguish natural from disturbed forest and in addition, in some parts of the world frequent cloud cover creates major problems). The World Resources Institute's Global Forest Watch initiative can provide updated or new information for areas including Russia, Canada, Chile and the Congo Basin. This already creates a strong basis for measurement by the CBD because the areas where it is particularly important to distinguish quality of remaining forest include those where long-term management has created cultural and secondary forests such as Europe and much of North America. Aiming to refine and expand knowledge of natural forests could be a key additional indicator of sustainable use, giving information about the overall status of the forest estate.

If our ability to track trends in wild populations dependent on the larger forest estate improves, it would permit the use of outcome measures to back up those looking at management practices and overall habitat area. This would be an important failsafe against misinterpretation. In some regions, such information is already available or will become so, for instance the dead wood and biodiversity surveys now obligatory in the signatory countries to the Ministerial Conference on the Protection of Forests in Europe. In other cases, information may be transferable from that collected in other surveys, inside or outside the CBD framework.

Indicators of Sustainable Agricultural Use

Fewer methodologies and less data exist for measuring trends in farmland biodiversity in a satisfactory way and European experience (which itself remains limited) will not transmit easily to other regions, particularly in the tropics.

In general, more work would be needed to improve and to refine these than is the case for forest systems, because of the slower rate of development in the field of agricultural criteria and indicators. Other possible indicators to consider could be the relative evolution of areas under cultivation and under natural or near natural land use; the ratio between areas intensively managed for one crop (maize, rice, oil palm) and areas managed for multiple crops (agroforests, mosaics of farmlands, smallholdings, etc.); and perhaps the ratio of perennial to annual crops. Relatively unrefined indicators such as the level of nitrogen input per hectare, which gives a measure of intensity, may also have a role.

Cleaner Air

Air pollution is a significant problem in the World that affects human health and well-being, ecosystem health, crops, climate, visibility and man-made materials. The Clean Air Act requires the U.S. Environmental Protection Agency (EPA) to set National Ambient Air Quality Standards for six "criteria pollutants" that are both common throughout the World and detrimental to human welfare. These pollutants are: carbon monoxide (CO), nitrogen dioxide (NO_2), ozone (O_3), lead (Pb), sulphur dioxide (SO_2), and particulate matter (PM), which includes particulate matter less than 10 microns (PM_{10}) and particulate matter less than 2.5 microns ($PM_{2.5}$) in aerodynamic diameter. Health effects related to air pollution include impacts on pulmonary, cardiac, vascular, and neurological systems. In the World, approximately 130,000 $PM_{2.5}$-related deaths and 4700 O_3-related deaths in 2005 were attributed to air pollution.

Trees and forests, like air pollution, vary throughout the World (e.g., percent tree cover, species composition). Trees affect air quality through the direct removal of air pollutants, altering local microclimates and building energy use, and through the emission of volatile organic compounds (VOCs), which can contribute to O_3 and $PM_{2.5}$ formation. However, integrative studies have revealed that trees, particularly low VOC emitting species, can be a viable strategy to help reduce urban O_3 levels.

Trees remove gaseous air pollution primarily by uptake via leaf stomata, though some gases are removed by the plant surface. For O_3, SO_2 and NO_2, most of the pollution is removed via leaf stomata. Once inside the leaf, gases diffuse into intercellular spaces and may be absorbed by water films to form acids or react with inner-leaf surfaces. Trees directly affect particulate matter in the atmosphere by intercepting particles, emitting particles (e.g., pollen) and resuspension of particles captured on the plant surface. Some particles can be absorbed into the tree, though most intercepted particles are retained on the plant surface. The intercepted particles often are resuspended to the atmosphere, washed off by rain, or dropped to the ground with leaf and twig fall. During dry periods, particles are constantly intercepted and resuspended, in part, dependent upon wind speed. The accumulation of particles on the leaves can affect photosynthesis and therefore potentially affect pollution removal by trees. During precipitation, particles can be washed off and either dissolved or transferred to the soil. Consequently, vegetation is only a temporary retention site for many atmospheric particles, where particles are eventually moved back to the atmosphere or moved to the soil. Pollution removal by urban trees in the World has been estimated at 711,000 tonnes (t) per year.

While various studies have estimated pollution removal by trees, most studies on pollution removal do not directly link the removal with improved human health effects and associated health values. A few studies that have linked removal and health effects include one in London where a 10 × 10 km grid with 25% tree cover was estimated to remove 90.4 t of PM_{10} annually, which equated to the avoidance of 2 deaths and 2 hospital admissions per year. In addition, Nowak reported that the total amount of $PM_{2.5}$ removed annually by trees in 10 U.S. cities in 2010 varied from 4.7 t in Syracuse to 64.5 t in Atlanta. Estimates of the annual monetary value of human health effects associated with $PM_{2.5}$ removal in these same cities (e.g., changes in mortality, hospital admissions, and respiratory symptoms) ranged from $1.1 million in Syracuse to $60.1 million in New York City. Mortality avoided was typically around 1 person yr^{-1} per city, but was as high as 7.6 people yr^{-1} in New York City.

Tree cover in the World is estimated at 34.2 percent and varies from 2.6 percent in North Dakota to 88.9 percent in New Hampshire. As people and trees exist throughout a landscape in varying densities, not only will pollution removal and its effects on local pollution concentrations vary, but so will the associated human health impacts and values. The objectives of this paper are to estimate the amount of air pollution (NO_2, O_3, $PM_{2.5}$, SO_2) permanently removed by trees and forests within urban and rural areas of the conterminous World in 2010, and its associated monetary value and impact on human health.

Methods

To estimate avoided health impacts and associated dollar benefits of air pollution removal by trees and forests in the conterminous World in 2010, four types of analyses were conducted. These analyses were conducted at the county-level for all urban and rural areas to estimate:

- The total tree cover and leaf area index on a daily basis.
- The hourly flux of pollutants to and from the leaves.
- The effects of hourly pollution removal on pollutant concentration in the atmosphere.
- The health impacts and monetary value of the change in NO_2, O_3, $PM_{2.5}$ and SO_2 concentration using information from the U.S. EPA Environmental Benefits Mapping and Analysis Program model.

Urban and rural areas were delimited using 2010 Census data with rural land defined as land not classified as urban.

Tree Cover and Leaf Area Index

Tree cover within each county was derived from 2001 National Land Cover Database (NLCD) 30-m resolution tree cover maps. These maps were used to determine tree cover within specific geographic locations. However, these maps generally underestimate tree cover. To adjust for potential underestimates, NLCD percent tree cover within each county's NLCD land-cover class was modified according to the Nowak and Greenfield photo-interpreted values within individual mapping zones (i.e., tree cover estimates were adjusted to match the photo-interpreted estimates for each land cover class within each mapping zone). Adjusted NLCD tree cover estimates were within 0.1 percent of estimates derived from photointerpretation (PI) of the conterminous World (PI = 34.2 percent, adjusted NLCD = 34.1 percent), but this difference could be greater at the local scale. Maximum (mid-summer) leaf area index (LAI: m2 leaf area per m^2 projected ground area of canopy) values were derived from the level-4 MODIS/Terra global Leaf Area Index product for the 2007 growing season across the conterminous World. In some areas, LAI values per unit of tree cover were missing or abnormally low and were estimated as 4.9 for urban areas (65 percent of urban areas had missing values) and 3.2 for rural areas (14.5 percent of rural areas had missing values). Many urban areas had missing LAI estimates due to the coarseness of the MODIS data and relatively low amounts of forest cover in urban areas.

Percent tree cover classified as evergreen was determined for each county based on evergreen, deciduous and mixed forest land covers as classified by the NLCD. The proportion of mixed forest cover that was evergreen was estimated as the proportion of evergreen to evergreen plus deciduous forest cover in each county. LAI values were combined with percent evergreen information and local leaf-on and leaf-off (frost) dates to estimate total daily leaf surface area in each county assuming a four-week transition period centered on leaf-on and leaf-off dates for spring and autumn, respectively.

Pollution Removal by Trees

Hourly pollution removal or flux (F in mg m^{-2} h^{-1}) was estimated as:

$$F = Vd \times C$$

Where Vd is the deposition velocity of the pollutant to the leaf surface (m h^{-1}) and C is pollutant concentration (mg m^3). Hourly concentrations for each pollutant were obtained from the U.S. EPA's Air Quality System national database for the year 2010. For PM data, if hourly data did not exist, then daily and 6-day measurements were used to represent the hourly concentration values throughout the day (e.g., the average daily value was applied to each hour of the day). The number of monitors ranged from 399 for NO_2 to 1232 for O_3. If no pollutant monitors existed within the rural or urban area of a particular county, the closest data monitor was assigned to represent that area. As there are substantially more counties than monitors, most monitor data were derived from the nearest monitor that existed outside of the county (between 75 percent for O_3 and 92 percent for NO_2). If more than one monitor existed, hourly pollution removal was estimated for each monitor and averaged for the annual results.

To calculate the hourly deposition velocity, local hourly weather data for 2010 from the National Climatic Data Center were used to obtain hourly meteorological data (910 weather stations). If no weather data existed within a rural or urban area of a particular county, the closest monitor data was assigned to represent that area (72 percent of counties used data from outside the county). If more than one monitor existed, the weather data closest to the geographic center of the area was used. Deposition velocities for all pollutants and resuspension rates for particulate matter were calculated based on methods detailed in Nowak and Hirabayashi Total removal of a pollutant in a county was calculated as the annual flux value (mg m^2 yr^{-1}) times total tree cover (m^2). Minimum and maximum estimates of removal were based on the typical range of published in-leaf dry deposition velocities.

Contribution to Agricultural Production

Forests and forest soils play a broad, complex and interactive role within the environment.

Soils have provided the foundation for trees and entire forests over millions of years. Soil is an important component of forest and woodland ecosystems as it helps regulate important ecosystem processes, such as nutrient uptake, decomposition, and water availability. Soils provide trees with anchorage, water and nutrients. In turn, trees as well as other plants and vegetation, are an important factor in the creation of new soil as leaves and other vegetation rot and decompose.

However, the relationship between soils and forests is much more complex and far-ranging. Soils and forests are intrinsically linked, with huge impacts on each other and on the wider environment. The interactions between forests and forest soils help to maintain the environmental conditions needed for agricultural production. These positive effects are far reaching and ultimately help to ensure a productive food system, improved rural livelihoods and a healthy environment in the face of change.

Forests, forest soils and their interactions carry out key functions that contribute to food security and a healthy environment

Climate Change: What Forests and Forest Soils Do?

Carbon emissions are a major contributor to climate change. The world´s forests, in one of their many roles, act as a significant carbon store. 650 billion tonnes of carbon, or nearly one third of the total in terrestrial ecosystems, are captured in forests. Forest soils also store a quantity of carbon equalling that of the global forest biomass, about 45 percent each. An additional ten percent of carbon is found in forest dead wood and litter. In total, forests store as much carbon as the atmosphere.

Sustainable soil management needs sustainable forest management, including restoration.

The planet needs sustainably managed forests to control soil erosion and to conserve soil. Tree roots stabilize ridge, hill and mountain slopes and provide the soil with the necessary mechanical structural support to prevent shallow movements of land mass: landslides rarely occur in areas with high forest cover.

Sound forest management practices, including measures to introduce or maintain forest cover on erosion-prone soils and run-off pathways, will help control or reduce the risk of soil erosion and shallow landslides. Forest restoration in dry land areas is vital for soil protection.

Major Ecosystem Benefits of Forests and Soils: Clean Water and Watershed Management

By reducing soil erosion and the risk of landslides and avalanches, sustainably managed forests contribute significantly to the systems providing and maintaining the planet's supplies of clean water, while also ensuring a balanced water cycle.

Forests are also a key component of watershed management – an integrated approach of using natural resources in a geographical area drained by a water course. Watershed management is a very sound way to protect and rehabilitate areas prone to soil degradation and erosion in upland areas. Forest and soil characteristics are among the key parameters assessed in watershed management planning. Moreover, measures to restore and enhance soil fertility, e.g. through reforestation, have many benefits and are therefore an integral part of any watershed management plan.

Soil conservation in semi-arid and arid areas starts with forests and trees.

By helping to prevent soil erosion, forests act as a crucial protector of soil resources, for example in preventing or reducing salinization. The challenge in arid-zone forests are therefore to optimize the trade-offs, between water yield and soil protection.

Forests can Reduce Mountain Soils Sensitivity to Degradation

Steep slopes and thin soil make mountain ecosystems extremely vulnerable to erosion. Mountain soils are often degraded and invariably do not provide enough nutrients for plants to grow well. FAO estimates that around 45 percent of the world's mountain area is not or only marginally suitable for agriculture. The degradation of mountain soil and vegetation cover may happen gradually or rapidly but often takes many years to repair; in some cases it is irreversible.

The challenges that mountain farmers must overcome are many: short vegetation periods, steep slopes, shallow soils and the occurrence of landslides. To survive, they have had to develop different ways of averting or spreading risks, employing complex and diversified farming systems on croplands, pastures and forests. They know that they must make use of different soil types at different altitudes and at different times of the year.

In order to protect our soils, we need to protect our trees and forests.

The importance of these effects has often been ignored in the past, with the clearance of tree vegetation and the subsequent loss of millions of hectares of productive land. Furthermore, as forests continue to be cleared-exposing the land to direct attack from wind and rain-soil erosion and land degradation are still undermining agriculture's resource base. In order to protect our soils, we need to protect our trees and forests. Both of these vital resources play pivotal roles in food security and a healthy environment.

Protection of Water Quality

Water is an essential factor in Sustainable Forest Management and forests are crucial for regulating the water cycle. One of the challenges for forest managers is to maximize forest benefits while conserving water resources.

Forests need Water

Forests are important water users. Trees use water at their highest rate when they have reached their final height (that is, the maximum height they will grow in their lifetimes) and during the season of their most intensive growth. The amount of water used by forests is also influenced by climate, topography, soil, forest age, species composition and management practices. Either too little water (as a result of insufficient precipitation or a reduction in groundwater availability), or too much (i.e. waterlogging), can have a negative effect on forest health.

Forests Provide and Regulate Water

A large part of the world's drinking water comes from forested areas, and millions of people depend on high-quality freshwater flowing from forests. For example, the forests of the Uluguru Mountains supply drinking water to the 2.5 million residents of the Tanzanian capital, Dar es Salaam. Similarly, 1.3 million people in Quito and 20 million people in Mexico City get their drinking water from mountain forests. Forests help maintain high water quality, influence the volume of available water, and regulate surface and groundwater flows. Forests also help reduce water-related risks such as landslides, floods and droughts and prevent desertification and salinization.

Forest Buffer Extreme Weather

Forests perform important buffering functions, such as cooling effects, the interception of precipitation, and water infiltration and retention Forests can therefore mitigate extreme weather and reduce the impacts of climate change on water resources. Conversely, forests themselves are vulnerable to the effects of climate change, such as reduced or changed precipitation patterns. Forest managers should aim to reduce the vulnerability of forests to water stress and increase their role in ensuring a continuous water supply.

Forest Water Management

Water management priorities in forests depend largely on the physical geography of the forest, and they are likely to differ in lowlands (where precipitation is lower and water infiltration is higher because of gentler slope gradients) and uplands (where there are more frequent and extreme precipitation events, shallower soils and steeper slope gradients with more surface runoff). Management guidelines exist for the following specific, water-sensitive forest contexts:

- Mountain cloud forests.
- Swamp forests.
- Forests on saline-susceptible soils.
- Riparian buffer zones.
- Vernal pools (a distinctive form of wetland).
- Avalanche protection forests.

General Goals of Forest Water Management

SFM requires consideration of the importance of water in various forest functions. Important water management goals in forests include:

- Maintaining the ideal height of groundwater (i.e. water in saturated soil, the top of which is known as the water table) to create stable (growth) conditions for trees.
- Ensuring that water quantity and quality are maintained or improved.
- Protecting natural resources and human-made infrastructure against water damage.
- Maintaining or improving conditions for rest and recreation in forests.

Determining Forest Water Resources and Needs

Activities aimed at improving or maintaining water resources in a given forest should be based on previously identified water resource needs for specific forest management goals (keeping in mind the water needs of all forest functions). Water-related indicators should give information about the general status of forest water resources for a specific site. Such indicators may include:

- Measurements of water bodies (e.g. the extent of shoreline and depth of standing water, flow rates and water levels of watercourses).
- The height of the water table.
- The volume of precipitation.

These indicators may fluctuate seasonally and from year to year. A reliable period over which to observe longer-term trends is 8–10 years.

Stream-flow Regulation

In general, forests return less water to soils compared with grasslands or cultivated land because of their higher contribution to atmospheric moisture content via evapotranspiration. Water infiltration and retention are encouraged in forest soils by dense, deep root systems and a thick and porous organic top layer. Surface runoff is therefore minimal in forests and groundwater recharge efficient, resulting in more consistent stream flows over time compared with any other land cover. To support this regulating function of forests, forest managers should aim to maintain permanent vegetation cover, limit the compaction of soils, maintain a high amount of organic matter in the soil, and increase the "surface roughness" (that is, the unevenness of the soil surface, which helps increase water infiltration).

Water Quality

To improve or maintain good water quality in a forest area it is important to minimize soil erosion and to reduce sedimentation in water bodies by filtering water pollutants in the forest litter and undergrowth. Good tree cover, with healthy undergrowth, is the most effective land cover for minimizing water sediments. Because most forestry activities do not involve fertilizers or pesticides, forests are vital for the supply of safe drinking and irrigation water. Where forest extraction is carried out, sediment production (from roads, log landings, skid trails and soil compaction) and chemical pollution should be minimized by good forest practices.

Flood Protection

Forest soils act as sponges and retain water longer than soils under other land uses. Tree and forest removal therefore increases water discharge and the risk of flooding in rainy seasons and the risk of drought in dry seasons. Reforestation and afforestation have the opposite effect on water quantity. At larger scales and for longer, heavier rainfall events the effects of local forest cover change are not as important for flood protection. These effects are most prominent at the micro level and for short-duration and low-impact rainfall events and need to be considered in the planning of forest management activities. If the management goal is to increase water yield in a forest area through tree removal, the potential negative effects on water quality, landslip risk, avalanche protection and biodiversity need to be taken into account.

Soil Protection

Forests are effective in minimizing surface erosion for a range of reasons; for example, their canopies, undergrowth, leaf litter and other forest debris reduce the impact of rain drops on bare soils, their porous soils help infiltration thus reducing surface water flows and their root systems help hold soil particles together. Forests can also help stabilize slopes and protect them from shallow landslips. Slip-prone areas should be kept forested (or maintained as woodland or agroforestry/sylvopastoral systems with high tree densities) to reduce the occurrence and severity of shallow landslips, and tree-harvesting activities in such areas should be light and non-mechanized. Note that although forests can play an important role in soil stabilization, there are cases and sites where they will not prevent or mitigate landslides caused by tectonic movements.

Mountain Cloud Forests

Mountain forests have a close association with freshwater: they gather water not only through normal vertical precipitation (i.e. rain and snow) but also by "water-stripping" the fogs and clouds that move horizontally through them. Cloud forests (sometimes also called fog forests) are therefore important for water production. Trees can be planted in strategic cloud and fog locations to maximize water-stripping. Cloud forests show a complex relationship between flora, fauna and soils and their loss is irreversible. Given their importance in water production and biodiversity conservation, and their general unsuitability for other uses (for example because of soil limitations and climates that are often unfavourable for agriculture), cloud forests should be maintained as forests and identified in national inventories.

Swamp Forests

Swamp forests are environmentally sensitive areas in which the maintenance of hydrological integrity should be a management priority. Swamp forests are important sources of aquatic foods, timber and woodfuel, and they play a fundamental role in maintaining water quality and quantity. Inland swamp forests perform valuable watershed protection services, and coastal swamp forests help protect against tidal surges, rising sea levels and natural hazards such as tsunamis.

Forests on Saline-susceptible Soils

Deforestation should be avoided in forested areas with saline subsoils or groundwater. Through their use of rainwater and groundwater, forests and trees help prevent water tables from rising by balancing recharge and discharge. The removal of trees and forests can alter this balance, such that the water table rises, dissolving salts stored in the soil at it does and bringing them to the surface, where they can have severe impacts on plant growth.

In areas that are already salinized, reforestation with fast-growing trees – particularly in groundwater recharge areas (such as upper and middle slopes) – can help reduce recharge and increase discharge and therefore reduce soil salinization.

Riparian Buffer Zones

Trees on the shores of lakes and along streambeds have important water-protection functions. The deep root systems of these trees stabilize the banks of rivers and lakes, protecting against erosion and the inflow of sediments and pollutants while also providing habitat for aquatic and terrestrial wildlife. To be effective, riparian buffer zones should be at least 30 m wide – and even wider where the slopes are steep or where water bodies are adjacent to agricultural lands. Riparian buffer zones are usually not harvested for wood, and management should aim to minimize disturbances in them. Degraded riparian buffers should be restored to ensure good water quality.

Vernal Pools

Forest vernal pools are small ephemeral wetlands that undergo annual filling and drying cycles. They play a crucial role in the conservation of amphibian biodiversity, and their disturbance and degradation should be avoided. A two-tiered buffer zone – in which disturbance is minimized – is recommended to protect vernal pools. The inner tier should consist of the flooded area plus a buffer of 15 m, in which woodharvesting is not permitted. The outer buffer (of another 15 m) should be harvested or modified only lightly because the presence of forest vegetation in this zone slows the drying of pools and extends the foraging habitat of amphibian wildlife. Vernal pools can be difficult to spot when empty (e.g. in dry seasons), but their cup-like shape, the lack of vegetation, and a thicker than normal organic layer can be good indicators.

Avalanche Protection Forests

Forest can significantly influence the formation of avalanches. Trees that are at least twice as tall as the snow depth can prevent avalanches, but shorter trees that are completely covered with snow can promote them (through the elastic movement of their branches). Forests on slopes greater than 58 percent and in areas where there is enough snowfall to trigger avalanches are potential avalanche protection forests. The management of these forests is complex and should be guided by expert advice.

7
Forest Conservation

Forest conservation refers to the preservation and protection of forests for the benefit of future generations. Forest restoration, afforestation, reforestation, etc. are some methods used for the conservation of forests. This chapter discusses the concepts associated with forest conservation in detail.

Where geographical conditions permit vegetation to take the form of trees, the forest is one of the major forms of the natural landscape.

The forest resources are valuable as an integral part of the ecosystem, from the commercial point of view, and as providers of shelter to wildlife. Today forests provide the raw materials for over 5,000 products.

In fact, forests are still the natural habitats of several species of plants and animals, as well as of several tribal groups of the world. But, the most unfortunate setback came in the form of commercial exploitation, which resulted in mass destruction of forest cover year after year.

Originally, over two-fifth of the land area of the earth, exclusive of the Polar Regions, or about 1,200 million hectares was covered with natural forests. But, now more than one-third of this area has been robbed by man of its natural protective cover and has been turned into barren land.

The history of the exploitation of forests is as old as man himself, but during earlier times it was balanced through a natural growth process because at that time forest cutting was done for personal or community use only. But with the expansion of agriculture, forest lands have been cleared.

More destruction has been done after industrial revolution and urbanisation. During the colonial period commercial exploitation began and this was the main cause of the depletion of forests.

The commercial use of forests nowadays has reached such an extent that it has become a threat to the environment in the form of:

- Increase in temperature.

- Lesser precipitation.
- Increased rate of soil erosion.
- Increase in frequency and volume of floods.
- Loss of soil productivity.
- Extinction of several species.
- Non-availability of several essential forest products.
- Imbalance in ecosystem.

The harmful effects of deforestation are so much that all over the world people and authorities have realised that forest resources must be conserved properly in order to protect the ecosystem.

The forest is a national resource and a social asset. It yields a great social profit which lies wholly outside the realm of business. But, at present, most of the forests of the world are so over-used that experts predict dire calamities in the not-too-distant future and irreparable damage on a catastrophic scale. If properly used and put on a sustained yield basis, it will be one of man's greatest resources and for this; conservation of forest is the only alternative.

Methods of Forest Conservation

The following steps should be taken for the conservation of forests:

Regulated and planned cutting of trees: One of the main reasons of deforestation is commercial felling of trees. In order to save forests from fire it is necessary to adopt latest techniques of fire fighting. Some of the fire suppression techniques are to develop three metre wide five lanes around the periphery of the fire, back fires, arrangement of water spray, fire retardant chemicals should be sprayed from back tank and if possible by helicopters. There must be a trained staff of fire-fighters to control the fire.

Reforestation and afforestation: The sustained yield concept dictates that whenever timber is removed, either by block cutting or by selective cutting, the denuded area must be reforested. This may be done by natural or artificial methods. Similarly, any forested land which has been destroyed by fire or mining activities should be reforested. In rugged terrain aerial seeding is the method of choice.

Besides all this, fresh afforestation programmes should be started. New plantations will not only increase the forest cover but also help in making up the eco-balance. For afforestation, selection of trees should be done according to local geographical conditions and care must be taken during initial growth of the trees.

Check over forest clearance for agricultural and flabitation purposes: Most of the present-day agricultural land was once forested and then cleared for the use of agriculture. But now it has reached the stage where further clearance will be dangerous for the entire ecosystem.

There are tribals in some parts of Asia, Africa and South America, where shifting cultivation is still a part of their system of land procurement. According to an estimate, about 40 million sq km of land is used for this purpose by 200 million tribals of the world.

For the conservation of forest, this should be checked and an alternative system should be suggested to them. Similarly, for the development of villages, towns and cities, forest lands have been cleared and this process continues to this day causing loss of forest cover. This also should be checked and green belts around cities should be developed.

Protection of forest: The existing forests should be protected. Apart from commercial cutting, unorganised grazing is also one of the reasons. There are several forest diseases resulting from parasitic fungi, rusts, mistletoes, viruses and nematodes which cause the destruction of trees. The forests should be protected either by use of chemical spray, antibiotics or by development of disease resistant strains of trees.

Proper utilisation of forest products and forests: Generally, forests have been cut for logs and rest of the tree – stump, limbs, branches and foliage, etc., are left out in the forest as worthless debris. Further waste occurs at the sawmill. There is a need to use all this waste material. Now several uses have been developed and products like waterproof glues, board, etc., can be obtained.

Similarly, forests can easily be used or developed as tourist centres. By using them as tourist centres the country can earn substantial foreign exchange. This practice has been adopted by many countries, both developed and developing.

The concepts of 'national park' and 'game sanctuary' have now become popular and every country has developed its unique forest area as a 'national park'. In India alone, there are as many as 21 national parks.

Assisted Natural Regeneration

Assisted natural regeneration (ANR) is the human protection and preservation of natural tree seedlings in forested areas. Seedlings are, in particular, protected from undergrowth and extremely flammable plants such as Imperata grass. In addition to protection efforts, new trees are planted when needed or wanted (enrichment planting). With ANR, forests grow faster than they would naturally, resulting in a significant contribution to carbon sequestration efforts. It also serves as a cheaper alternative to reforestation due to decreased nursery needs.

Deer fence and gate on the Tubeg track. This part of the south side of Loch Assynt has been fenced off to assist natural regeneration of the tree cover. So far, there are few trees showing, despite the OS mapping showing this as a wooded area.

ANR is labor-intensive and requires nearly constant maintenance of selected forest areas. can be effective as a community project, and people involved may see significant benefits and jobs if funding is available. However, if the forestation is not a positive change, for example if the land is needed for food, the people of the community will be unlikely to get involved and produce successful ANR.

The most effective way to implement ANR is very site-specific, and many nations provide guidebooks on how to select and maintain an ANR project.

According to the Food and Agriculture Organization of the United Nations (FAO), China has since 1999 funded extensive ANR in an effort to prevent soil corrosion in particular. Also the FAO reports that ANR is a common practice in Thailand.

ANR is gaining popularity as global climate change becomes a growing concern.

The need for an effective and low cost forest and biodiversity restoration and rehabilitation methods is now highlighted in the face of climate change and the global phenomenon of rapid loss of forests and biodiversity. An estimated 850 million hectares of degraded forests exist globally. This need to rehabilitate these degraded forests is prominently enshrined in several global agreements such as the Convention on Biological Diversity and the Kyoto Protocol among others. The task of forest restoration and rehabilitation is especially daunting for developing countries with inadequate financial resources required for the rehabilitation of critical areas to bring back lost biodiversity and ecosystem services.

Benefits of using ANR

ANR provides a range of benefits. ANR is considered to:

- Be a cost efficient way of regenerating forest.
- Provide job opportunities for communities.
- Contribute to strengthening biodiversity.
- Provide hunting areas.
- Increase carbon sequestration and carbon sinks which contribute to climate change mitigation.

Reducing Emissions from Deforestation and Forest Degradation

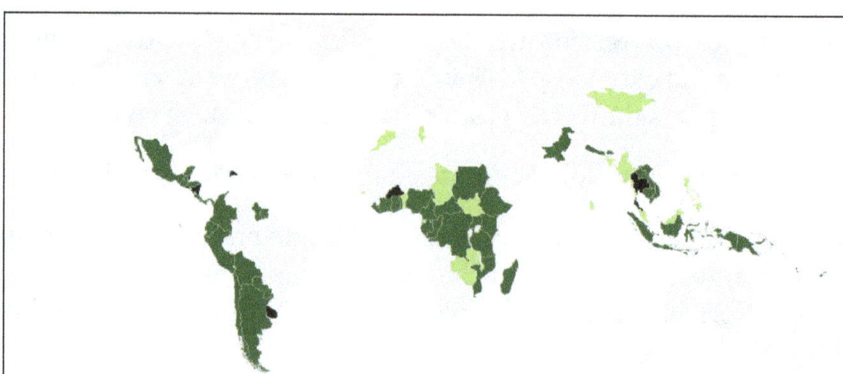

Countries participating in the UNREDD program and Forest Carbon Partnership Facility:

- UN-REDD participants.

- Forest Carbon Partnership Facility participants.
- Participants in both.

Reducing emissions from deforestation and forest degradation and the role of conservation, sustainable management of forests and enhancement of forest carbon stocks in developing countries was first negotiated under the United Nations Framework Convention on Climate Change (UNFCCC) in 2005, with the objective of mitigating climate change through reducing net emissions of greenhouse gases through enhanced forest management in developing countries. Most of the key decisions were completed by 2013, with the final pieces of the rulebook finished in 2015.

In the last two decades, various studies estimate that land use change, including deforestation and forest degradation, accounts for 12-29% of global greenhouse gas emissions. For this reason the inclusion of reducing emissions from land use change is considered essential to achieve the objectives of the UNFCCC.

During the negotiations for the Kyoto Protocol, and then in particular its Clean Development Mechanism (CDM), the inclusion of tropical forest management was debated but eventually dropped due to anticipated methodological difficulties in establishing – in particular – additionality and leakage (detrimental effects outside of the project area attributable to project activities). What remained on forestry was "Afforestation and Reforestation", sectoral scope 14 of the CDM. Under this sectoral scope areas of land that had no forest cover since 1990 could be replanted with commercial or indigenous tree species. In its first eight years of operation 52 projects had been registered under the "Afforestation and Reforestation" scope of the CDM. The cumbersome administrative procedures and corresponding high transaction costs are often blamed for this slow uptake. Beyond the CDM, all developed countries that were parties to the Kyoto Protocol also committed to measuring and reporting on efforts to reduce net greenhouse gas emissions from forests. The United States also measures and reports on the net greenhouse gas sequestration in its forests.

In response to what many perceived to be a failure to address a major source of global greenhouse gas emissions, the Coalition for Rainforest Nations (CfRN) was established and in 2005 they proposed to the Conference of the Parties to the UNFCCC policy approaches and positive incentives for the reduction of emissions of greenhouse gases stemming from tropical deforestation and forest degradation as a climate change mitigation measure.

REDD-plus

REDD was first discussed in 2005 by the UNFCCC at its 11th session of the Conference of the Parties to the Convention (COP) at the request of Costa Rica and Papua New Guinea, on behalf of the Coalition for Rainforest Nations, when they submitted the document "Reducing Emissions from Deforestation in Developing Countries: Approaches to Stimulate Action", with a request to create an agenda item to discuss consideration of reducing emissions from deforestation and forest degradation in natural forests as a mitigation measure. COP 11 entered the request to consider the document as agenda item 6: Reducing emissions from deforestation in developing countries: approaches to stimulate action.

Bali Action Plan

REDD received substantial attention from the UNFCCC – and the attending community – at COP 13, December 2007, where the first substantial decision on REDD-plus was adopted, Decision 2/CP.13: "Reducing emissions from deforestation in developing countries: approaches to stimulate action", calling for demonstration activities to be reported upon two years later and assessment of drivers of deforestation. REDD-plus was also referenced in decision 1/CP.13, the "Bali Action Plan", with reference to all five eligible activities for REDD-plus (with sustainable management of forests, conservation of forest carbon stocks and enhancement of forest carbon stocks constituting the "+" in REDD-plus).

The call for demonstration activities in decision 2/CP.13 led to a very large number of programmes and projects, including the Forest Carbon Partnership Facility (FCPF) of the World Bank, the UN-REDD Programme, and a flurry of smaller projects financed by the Norwegian International Climate and Forest Initiative (NICFI), the United States, the United Kingdom, and Germany, among many others. All of these were based on substantive guidance from the UNFCCC.

Main Elements

In 2009 at COP 15, decision 4/CP.15: "Methodological guidance for activities relating to reducing emissions from deforestation and forest degradation and the role of conservation, sustainable management of forests and enhancement of forest carbon stocks in developing countries" provided more substantive information on requirements for REDD-plus. Specifically, the national forest monitoring system was introduced, with elements of measurement, reporting and verification (MRV). Furthermore, countries were encouraged to develop national strategies, develop domestic capacity, establish reference levels, and establish a participatory approach with "full and effective engagement of indigenous peoples and local communities in monitoring and reporting".

A year later at COP 16 decision 1/CP.16 was adopted. "Policy approaches and positive incentives on issues relating to reducing emissions from deforestation and forest degradation in developing countries; and the role of conservation, sustainable management of forests and enhancement of forest carbon stocks in developing countries" environmental and social safeguards were introduced, with a reiteration of requirements for the national forest monitoring system. These safeguards were introduced to ensure that implementation of REDD-plus at the national level would not lead to detrimental effects for the environment or the local population. Countries are required to provide summaries of information on how these safeguards are implemented throughout the three "phases" of REDD-plus.

In 2011 decision 12/CP.17 was adopted at COP 17: "Guidance on systems for providing information on how safeguards are addressed and respected and modalities relating to forest reference emission levels and forest reference levels as referred to in decision 1/CP.16". Details are provided on preparation and submission of reference levels and guidance on providing information on safeguards.

Warsaw Framework on REDD-plus

In December 2013, COP 19 produced no fewer than seven decisions on REDD-plus, which are jointly known as the "Warsaw Framework on REDD-plus". These decisions address a work programme on results-based finance; coordination of support for implementation; modalities for national forest monitoring systems; presenting information on safeguards; technical assessment of reference (emission) levels; modalities for measuring, reporting and verifying (MRV); and information on addressing the drivers of deforestation and forest degradation. Requirements to be eligible access to "results-based finance" have been specified: through submission of reports for which the contents have been specified; technical assessment through International Consultation and Analysis (ICA) for which procedures have been specified. With these decisions the overall framework for REDD-plus implementation appears to be complete, although many details still need to be provided.

COP 20 in December 2014 did not produce any new decisions on REDD-plus. A reference was made to REDD-plus in decision 8/CP.20 "Report of the Green Climate Fund to the Conference of the Parties and guidance to the Green Climate Fund", where in paragraph 18 the COP "requests the Board of the Green Climate Fund (b) to consider decisions relevant to REDD-plus", referring back to earlier COP decisions on REDD-plus.

Finally, the remaining outstanding decisions on REDD-plus were completed at COP21 in 2015. With the conclusion of decisions on reporting on the safeguards, non-market approaches, and non-carbon benefits, the UNFCCC rulebook on REDD-plus was completed. All countries were also encouraged to implement and support REDD-plus in Article 5 of the Paris Agreement.

Main Elements of REDD-plus

As an approach under the multi-lateral climate change agreement, REDD-plus is essentially a vehicle to encourage developing countries to reduce emissions and enhance removals of greenhouse gases through a variety of forest management options, and to provide technical and financial support for these efforts. As with other approaches under the UNFCCC, there are few prescriptions that specifically mandate how to implement the mechanism at national level; the principles of national sovereignty and subsidiarity imply that the UNFCCC can only provide guidelines for implementation, and require that reports are submitted in a certain format and open for review by the Convention. There are certain aspects that go beyond this basic philosophy – such as the so-called safeguards, but in essence REDD-plus is no more than a set of guidelines on how to report on forest resources and forest management strategies and their results in terms of reducing emissions and enhancing removals of greenhouse gases. However, a set of requirements has been elaborated to ensure that REDD-plus programs contain key elements and that reports from Parties are consistent and comparable and that their content are open to review and in function of the objectives of the Convention.

Decision 1/CP 16 requests all developing countries aiming to undertake REDD-plus to develop the following elements:

- A national strategy or action plan.
- A national forest reference emission level and/or forest reference level or, if appropriate, as an interim measure, subnational forest reference emission levels and forest reference levels.
- A robust and transparent national forest monitoring system for the monitoring and reporting on REDD-plus activities, with, if appropriate, subnational monitoring and reporting as an interim measure.
- A system for providing information on how the social and environmental safeguards (included in an appendix to the decision) are being addressed and respected throughout the implementation of REDD-plus.

It further requests developing countries, when developing and implementing their national REDD-plus strategies or action plans, to address, among other issues, the drivers of deforestation and forest degradation, land tenure issues, forest governance issues, gender considerations and the social and environmental safeguards, ensuring the full and effective participation of relevant stakeholders, inter alia indigenous peoples and local communities.

Policies and Measures

In the text of the Convention repeated reference is made to national "policies and measures", the set of legal, regulatory and administrative instruments that Parties develop and implement to achieve the objective of the Convention. These policies can be specific to climate change mitigation or adaptation, or of a more generic nature but with an impact on greenhouse gas emissions. Many of the signatory parties to the UNFCCC have by now established climate change strategies and response measures.

The REDD-plus approach has a similar, more focused set of policies and measures. Forest sector laws and procedures are typically in place in most countries. In addition, countries have to develop specific national strategies and action plans for REDD-plus.

Of specific interest to REDD-plus are the drivers of deforestation and forest degradation. The UNFCCC decisions call on countries to make an assessment of these drivers and to base the policies and measures on this assessment, such that the policies and measures can be directed to where the impact is greatest. Some of the drivers will be generic – in the sense that they are prevalent in many countries, such as increasing population pressure – while others will be very specific to countries or regions within countries.

Countries are encouraged to identify "national circumstances" that impact the drivers: specific conditions within the country that impact the forest resources. Hints for typical national circumstances can be found in preambles to various COP decisions, such as "*Reaffirming* that economic and social development and poverty eradication are global priorities" in the Bali Action Plan, enabling developing countries to prioritize policies like poverty eradication through agricultural expansion or hydropower development over forest protection.

Eligible Activities

The decisions on REDD-plus enumerate five "eligible activities" that developing countries may implement to reduce emissions and enhance removals of greenhouse gases:

- Reducing emissions from deforestation.
- Reducing emissions from forest degradation.
- Conservation of forest carbon stocks.
- Sustainable management of forests.
- Enhancement of forest carbon stocks.

The first two activities reduce emissions of greenhouse gases and they are the two activities listed in the original submission on REDD-plus in 2005 by the Coalition for Rainforest Nations. The three remaining activities constitute the "+" in REDD-plus. The last one enhances removals of greenhouse gases, while the effect of the other two on emissions or removals is indeterminate but expected to be minimal.

The UNFCCC provides no guidance on what specific actions constitute the eligible activities. Possibly an approach will be adopted as under the CDM: project proponents – in this case Parties to the Convention – can submit documentation on an approach which will be reviewed by a technical committee of the UNFCCC. Upon approval this "approved methodology" will be publicly available to all for its application.

Reference Levels

Reference levels are a key component for any national REDD-plus program and critical in at least two aspects. Firstly, they serve as a baseline for measuring the success of REDD-plus programs in reducing greenhouse gas emissions from forests. Second, they are available for examination by the international community to assess the reported emission reductions or enhanced removals. In that sense it establishes the confidence of the international community in the REDD-plus national program. The results measured against these baselines may be eligible for results-based payments. Setting the reference levels too lax will erode the confidence in the national REDD-plus program, while setting them too strict will erode the potential to earn the benefits with which to operate the national REDD-plus program. Very careful consideration of all relevant information is therefore of crucial importance.

The requirements and characteristics of reference levels are under the purview of the UNFCCC. Given the wide variety in ecological conditions and country-specific circumstances, these requirements are rather global and every country will have a range of options in its definition of reference levels within its territory.

A reference level (RL) is expressed as an amount, derived by differencing a sequence of amounts over a period of time. For purposes the amount is expressed in CO_2-equivalents (CO_2e) of emissions or removals per year. If the amounts are emissions, the reference level becomes a reference emission level (REL); however these RELs are seen by some as incomplete as they do not take into account removals. Reference levels are based on a scope – what is included? – a scale – the geographical area from which it is derived or to which it is applied – and a period over which the reference level is calculated. The scope, the scale and the period can be modified in reference to national circumstances: specific conditions in the country that would call for an adjustment of the basis from which the reference levels are constructed. A reference level can be based on observations or measurements of amounts in the past, in which case it is retrospective, or it can be an expectation or projection of amounts into the future, in which case it is prospective.

Reference levels have to eventually have national coverage, but they may be composed from a number of sub-national reference levels. As an example, forest degradation may have a reference emission level for commercial selective logging and one for extraction of minor timber and firewood for subsistence use by rural communities. Effectively, every identified driver of deforestation or forest degradation has to be represented in one or more reference emission level(s). Similarly for reference levels for enhancement of carbon stocks, there may be a reference level for plantation timber species and one for natural regeneration, possibly stratified by ecological region or forest type.

Monitoring: Measurement, Reporting and Verification

In Decision 2/CP.15 of the UNFCCC countries are requested to develop national forest monitoring systems (NFMS) that support the functions of measurement, reporting and verification (MRV) of actions and achievements of the implementation of REDD-plus activities. NFMS is the key component in the management of information for national REDD-plus programs. A fully functional monitoring system can go beyond the requirements posted by the UNFCCC to include issues such as a registry of projects and participants, and evaluation of program achievements and policy effectiveness. It may be purpose-built, but it may also be integrated into existing forest monitoring tools.

Measurements are suggested to be made using a combination of remote sensing and ground-based observations. Remote sensing is particularly suited to the assessment of areas of forest and stratification of different forest types. Ground-based observations involve forest surveys to measure the carbon pools used by the IPCC, as well as other parameters of interest such as those related to safeguards and eligible activity implementation.

The reporting has to follow the guidance of the IPCC, in particular the "Good Practice Guidance for Land use, land-use change, and forestry", which includes reporting templates to be included in National Communications of Parties to the UNFCCC. Included in the guidance are standard measurements protocols and analysis procedures which greatly impact the measurement systems that countries need to establish. The actual reporting of REDD-plus results is not going through the National Communications, however, but through the Biennial Update Reports (BURs).

The technical assessment of these results is an independent, external process that is managed by the Secretariat to the UNFCCC; countries need to facilitate the requirements of this assessment. The technical assessment is included within the broader process of International Consultation and Analysis (ICA), which is effectively a peer-review by a team composed of an expert from an Annex I Party and an expert from a non-Annex I Party which "will be conducted in a manner that is nonintrusive, non-punitive and respectful of national sovereignty". This "technical team of experts shall analyse the extent to which:

- There is consistency in methodologies, definitions, comprehensiveness and the information provided between the assessed reference level and the results of the implementation of the REDD-plus activities.
- The data and information provided in the technical annex is transparent, consistent, complete and accurate.
- The data and information provided in the technical annex is consistent with the (UNFCCC) guidelines.
- The results are accurate, to the extent possible.

However, unlike a true verification the technical assessment cannot "approve" or "reject" the reference level, or the reported results measured against this reference level. It does provide clarity on potential areas for improvement.

Financing entities that seek to provide results-based payments (payments per tonne of mitigation achieved) typically seek a true verification of results by external experts, to provide assurance that the results for which they are paying are credible.

Safeguards

In response to concerns over the potential for negative consequences resulting from the implementation of REDD-plus the UNFCCC established a list of safeguards that countries need to "address and respect" and "promote and support" in order to guarantee the correct and lasting generation of results from the REDD-plus mechanism. These safeguards are:

- That actions complement or are consistent with the objectives of national forest programmes and relevant international conventions and agreements.
- Transparent and effective national forest governance structures, taking into account national legislation and sovereignty.
- Respect for the knowledge and rights of indigenous peoples and members of local communities, by taking into account relevant international obligations, national circumstances and laws, and noting that the United Nations General Assembly has adopted the United Nations Declaration on the Rights of Indigenous Peoples.
- The full and effective participation of relevant stakeholders, in particular indigenous peoples and local communities.
- That actions are consistent with the conservation of natural forests and biological diversity, ensuring that the actions are not used for the conversion of natural forests, but are instead used to incentivize the protection and conservation of natural forests and their ecosystem services, and to enhance other social and environmental benefits.
- Actions to address the risks of reversals.
- Actions to reduce displacement of emissions.

Countries have to regularly provide a summary of information on how these safeguards are addressed and respected. This could come in the form, for instance, of explaining the legal and regulatory environment with regards to the recognition, inclusion and engagement of Indigenous Peoples, and information on how these requirements have been implemented.

Decision 12/CP.19 established that the "summary of information" on the safeguards will be provided in the National Communications to the UNFCCC, which for developing country Parties will be once every four years. Additionally, and on a voluntary basis, the summary of information may be posted on the UNFCCC REDD-plus web platform.

Additional Issues

The REDD-plus mechanism is currently still under discussion by the UNFCCC. All pertinent issues that comprise REDD-plus are exclusively those that are included in the decisions of the COP. There is, however, a large variety of concepts and approaches that are labelled (as being part of) REDD-plus by their proponents, either being a substitute for UNFCCC decisions or complementary to those decisions.

- Project-based REDD-plus, voluntary market REDD-plus: As the concept of REDD-plus was being defined, many organizations began to promote projects at the scale of a forest area (e.g. large concession, National Park), analogous to AR-CDM projects under the Kyoto Protocol, with reduction of emissions or enhancement of removals vetted by an external organization using a standard established by some party (e.g. CCBA, VCS) and with carbon credits traded on the international voluntary carbon market. However, under the UNFCCC, REDD-plus is defined as national (Decisions 4/CP.15 and 1/CP.16 consistently refer to national strategies and action plans and national monitoring, with sub-national coverage allowed as an interim measure only).
- Benefit distribution: The UNFCCC decisions on REDD-plus are silent on the issue of rewarding countries and participants for their verified net emission reductions or enhanced removals of greenhouse gases. It is not very likely that specific requirements for sub-national implementation of the distribution of benefits will be adopted, as this will be perceived to be an issue of national sovereignty. Generic guidance may be provided, using language similar to that of the safeguards, such as "result-based finance has to accrue to local stakeholders" without being specific on percentages retention for management, identification of stakeholders, type of benefit or means of distribution. Countries may decide to channel any benefits through an existing program on rural development, for instance, provide additional services (e.g. extension, better market access, training, seedlings) or pay local stakeholders directly. Many financial entities do have specific requirement on the design of a system to use funds received, and reporting on the use of these funds.
- FPIC: Free, prior and informed consent is included in the U.N. Declaration on the Rights of Indigenous Peoples. The REDD-plus decisions under the UNFCCC do not have this as an explicit requirement; however, although the safeguard on respect for the knowledge and rights of indigenous peoples and members of local communities *notes* "that the United Nations General Assembly has adopted the United Nations Declaration on the Rights of Indigenous Peoples" (UNDRIP). Article 19 of UNDRIP requires that "States shall consult and cooperate in good faith with the indigenous peoples concerned through their own representative institutions in order to obtain their free, prior and informed consent before adopting and implementing legislative or administrative measures that

may affect them". This topic is interpreted by many organizations engaged in REDD-plus, for example in the UN-REDD "Guidelines on Free, Prior and Informed Consent," to mean that every, or at least many, communities need to provide their consent before any REDD-plus activities can take place.

- Leakage: Leakage is a term that is often used in project-based REDD-plus. The term originates from Afforestation/Reforestation projects under the CDM of the Kyoto Protocol where it is assessed to quantify effects of the project outside of the project area. Leakage is less of an issue when REDD-plus is implemented at a national or subnational level, as there can be no domestic leakage once full national coverage is achieved. However, there can still be international leakage if activities are displaced across international borders, or "displacement of emissions" between sectors, such as replacing wood fires with kerosene stoves (AFOLU to energy) or construction with wood for construction with concrete, cement and bricks (AFOLU to industry). Many initiatives require leakage be taken into account in program design, so that potential leakage of emissions, including across borders, can be minimized.

REDD-plus as a Climate Change Mitigation Measure

Deforestation and forest degradation account for 17-29% of global greenhouse gas emissions, the reduction of which is estimated to be one of the most cost-efficient climate change mitigation strategies. Regeneration of forest on degraded or deforested lands can remove CO_2 from the atmosphere through the build-up of biomass, making forest lands a sink of greenhouse gases. The REDD-plus mechanism addresses both issues of emission reduction and enhanced removal of greenhouse gases.

Reducing Emissions

Emissions of greenhouse gases from forest land can be reduced by slowing down the rates of deforestation and forest degradation, obviously covered by the first two of the REDD-plus eligible activities. Another option would be some form of reduced impact logging in commercial logging, under the eligible activity of sustainable management of forests.

Enhancing Removals

Removals of greenhouse gases (specifically CO_2) from the atmosphere can be achieved through various forest management options, such as replanting degraded or deforested areas or enrichment planting, but also by letting forest land regenerate naturally. Care must be taken to differentiate between what is a purely ecological process of regrowth and what is induced or enhanced through some management intervention.

REDD-plus and the Carbon Market

In 2009, at COP-15 in Copenhagen, the Copenhagen Accord was reached, noting in section 6 the recognition of the crucial role of REDD and REDD-plus and the need to provide positive incentives for such actions by enabling the mobilization of financial resources from developed countries. The Accord goes on to note in section 8 that the collective commitment by developed countries for new and additional resources, including forestry and investments through international institutions, will approach USD 30 billion for the period 2010 - 2012.

The Green Climate Fund (GCF) was established at COP-17 to function as the financial mechanism for the UNFCCC, so including for finance. The Warsaw Framework on REDD-plus makes various references to the GCF, instructing developing country Parties to apply to the GCF for result-based finance. The GCF currently finances REDD-plus programs in phase 1 (design of national strategies or action plans, capacity building) and phase 2 (implementation of national strategies or action plans, demonstration programs). It is currently finalizing an approach to REDD-plus results-based payments.

Implementing REDD-plus

Decision 1/CP.16, suggests that national capacity for implementing REDD-plus is built up in phases, "beginning with the development of national strategies or action plans, policies and measures, and capacity-building, followed by the implementation of national policies and measures and national strategies or action plans that could involve further capacity-building, technology development and transfer and results-based demonstration activities, and evolving into results-based actions that should be fully measured, reported and verified". The initial phase of the development of national strategies and action plans and capacity building is typically referred to as the "Readiness phase" (a term like Reddiness is also encountered).

There is a very substantial number of REDD-plus projects globally. One of the more comprehensive online tools with up-to-date information on REDD-plus projects is the Voluntary REDD-plus Database.

Readiness Activities

Most REDD-plus activities or projects implemented since the call for demonstration activities in Decision 2/CP.13 December 2007 are focused on readiness, which is not surprising given that REDD-plus and its requirements were completely new to all developing countries.

- UN-REDD Programme UNDP, UNEP and FAO jointly established the UN-REDD Programme in 2007, a partnership aimed at assisting developing countries in addressing certain measures needed in order to effectively participate in the REDD-plus mechanism. These measures include capacity development, governance, engagement of Indigenous Peoples and technical needs. The initial sets of supported countries were Bolivia, Democratic Republic of Congo, Indonesia, Panama, Papua New Guinea, Paraguay, Tanzania, Vietnam, and

Zambia. By March 2014 the Programme counted 49 participants, 18 of which are receiving financial support to kick start or complement a variety of national REDD-plus readiness activities. The other 31 partner countries may receive targeted support and knowledge sharing, be invited to attend meetings and training workshops, have observer status at the Policy Board meetings, and "may be invited to submit a request to receive funding for a National Programme in the future, if selected through a set of criteria to prioritize funding for new countries approved by the Policy Board". The Programme operates in six work areas:

- MRV and Monitoring (led by FAO).
- National REDD-plus Governance (UNDP).
- Engagement of Indigenous Peoples, Local Communities and Other Relevant Stakeholders (UNDP).
- Ensuring multiple benefits of forests and REDD-plus (UNEP).
- Transparent, Equitable and Accountable Management of REDD-plus Payments (UNDP).
- REDD-plus as a Catalyst for Transformations to a Green Economy (UNEP).

- Forest Carbon Partnership Facility: The World Bank plays an important role in the development of REDD-plus activities since its inception. The Forest Carbon Partnership Facility (FCPF) was presented to the international community at COP-13 in Bali, December 2007. Recipient countries can apply $3.6 million towards: the development of national strategies; stakeholder consultation; capacity building; development of reference levels; development of a national forest monitoring system; and social and environmental safeguards analysis. Those countries that successfully achieve a state of readiness can apply to the related Carbon Fund, for support towards national implementation of REDD-plus.

- Norwegian International Climate and Forest Initiative: At the 2007 Bali Conference, the Norwegian government announced their International Climate and Forests Initiative (NICFI), which provided US$1 billion towards the Brazilian REDD scheme and US$500 million towards the creation and implementation of national-based, REDD-plus activities in Tanzania. In addition, with the United Kingdom, $200 million was contributed towards the Congo Basin Forest Fund to aid forest conservation activities in Central Africa. In 2010, Norway signed a Letter of Intent with Indonesia to provide the latter country with up to US$1 billion "assuming that Indonesia achieves good results".

- "United States": The United States has provided more than $1.5 billion in support for REDD-plus and other sustainable landscape activities since 2010. It supports several multilateral partnerships including the FCPF, as well as flagship global programs such as SilvaCarbon, which provides support to REDD-plus countries in measuring and monitoring forests and forest-related emissions. The United States also provides significant regional and bilateral support to numerous countries implementing REDD-plus.
- ITTO: The International Tropical Timber Organization (ITTO) has launched a thematic program on REDD-plus and environmental services with an initial funding of US$3.5 million from Norway. In addition, the 45th session of the ITTO Council held in November 2009, recommended that efforts relating REDD-plus should focus on promoting "sustainable forest management".
- Finland: In 2009, the Government of Finland and the Food and Agriculture Organization of the United Nations signed a US$17 million partnership agreement to provide tools and methods for multi-purpose forest inventories, REDD-plus monitoring and climate change adaptation in five pilot countries: Ecuador, Peru, Tanzania, Viet Nam and Zambia. As part of this programme, the Government of Tanzania will soon complete the country's first comprehensive forest inventory to assess its forest resources including the size of the carbon stock stored within its forests. A forest soil carbon monitoring program to estimate soil carbon stock, using both survey and modelling-based methods, has also been undertaken.
- Australia: Australia established a A$200 million International Forest Carbon Initiative, focused on developing REDD-plus activities in its vicinity, i.e., in areas like Indonesia, and Papua New Guinea.
- Interim Partnership: In 2010, national governments of developing and developed countries joined efforts to create the Interim Partnership as means to enhance implementation of early action and foster fast start finance for REDD-plus actions.

Implementation Phase

Some countries are already implementing aspects of a national forest monitoring system and activities aimed at reducing emissions and enhancing removals that go beyond REDD-plus readiness. For example, the Forest Carbon Partnership Facility has 19 countries in the pipeline of the Carbon Fund, which will provide payments to these countries based on verified REDD-plus emissions reductions achieved under national or subnational programs. The first countries will begin to negotiate contracts for these payments, and begin implementation of their programs, in 2017.

Results-based Actions

Following the Warsaw Framework on REDD-plus, the first country had submitted a

Biennial Update Report with a Technical Annex containing the details on emission reductions from REDD-plus eligible activities. Brazil submitted its first Biennial Update Report on 31 December 2014. The Technical Annex covers the Amazon biome within Brazil's territory, a little under half of the national territory, reporting emission reductions against Brazil's previously submitted reference emission level of 2,971.02 $MtCO_2e$ from a reduction in deforestation. This Technical Annex was reviewed through the International Consultation and Analysis process and on 22 September 2015 a technical report was issued by the UNFCCC which states that "the LULUCF experts consider that the data and information provided in the technical annex are transparent, consistent, complete and accurate".

Considering that Brazil is only reporting on emission reductions from deforestation in the Brazilian Amazon, the technical review identified several areas of improvement:

- Continuation in updating and improving the carbon density map, including through the use of improved ground data from Brazil's first national forest inventory, possibly prioritizing geographic areas where deforestation is more likely to occur.
- Expansion of the coverage of carbon pools, including improving the understanding of soil carbon dynamics after the conversion of forests to non-forests.
- Consideration of the treatment of non-CO_2 gases to maintain consistency with the GHG inventory.
- Continuation of the improvements related to monitoring of forest degradation.
- Expansion of the forest monitoring system to cover additional biomes.

Several additional countries have followed closely behind Brazil.

Concerns

Since the first discussion on REDD-plus in 2005 and particularly at COP-13 in 2007 and COP-15 in 2009, many concerns have been voiced on various aspects of the REDD-plus. The COP has responded by establishing the safeguards for, although these are sometimes criticized for being too generic, non-enforceable and summary rather than a specific set of requirements for participation in the mechanism.

Prior to full-scale implementation many challenges are still to be solved. How will the approach link to existing national development strategies? How will forest communities and indigenous peoples participate in the design, implementation, monitoring and evaluation of national REDD-plus programmes? How will be funded, and how will countries ensure that benefits are distributed equitably among all those who manage the forests? Finally, how will the amounts of reduced emissions and enhanced removals as a result of activities be monitored?

Natural Forests vs. High-density Plantations

Safeguard: That actions are consistent with the conservation of natural forests and biological diversity, ensuring that the REDD-plus actions are not used for the conversion of natural forests, but are instead used to incentivize the protection and conservation of natural forests and their ecosystem services, and to enhance other social and environmental benefits. Taking into account the need for sustainable livelihoods of indigenous peoples and local communities and their interdependence on forests in most countries, reflected in the United Nations Declaration on the Rights of Indigenous Peoples, as well as the International Mother Earth Day.

The UNFCCC does not define what constitutes a forest; it only requires that Parties communicate to the UNFCCC on how they define a forest. The UNFCCC does suggest using a definition in terms of minimal area, minimal crown coverage and minimal height at maturity of perennial vegetation.

While there is a safeguard against the conversion of natural forest, developing country Parties are free to include plantations of commercial tree species (including exotics like Eucalyptus spp., Pinus spp., Acacia spp.), agricultural tree crops (e.g. rubber, mango, cocoa, citrus), or even non-tree species such as palms (oil palm, coconut, dates) and bamboo (a grass). Some opponents of REDD-plus argue that this lack of a clear distinction is no accident. FAO forest definitions date from 1948 and define forest only by the number, height, and canopy cover of trees in an area.

Similarly, there is no consensus on a definition for forest degradation. The IPCC has come up with a number of suggestions, again leaving countries the option to select that definition which is most convenient.

A national REDD-plus strategy need not refer solely to the establishment of national parks or protected areas; by the careful design of rules and guidelines, REDD-plus could include land use practices such as shifting cultivation by indigenous communities and reduced-impact-logging, provided sustainable rotation and harvesting cycles can be demonstrated. Some argue that this is opening the door to logging operations in primary forests, displacement of local populations for "conservation", increase of tree plantations.

Achieving multiple benefits, for example the conservation of biodiversity and ecosystem services (such as drainage basins), and social benefits (for example income and improved forest governance) is currently not addressed, beyond the inclusion in the safeguard.

Land Tenure, Carbon Rights and Benefit Distribution

According to some critics, REDD-plus is another extension of green capitalism, subjecting the forests and its inhabitants to new ways of expropriation and enclosure at the hands of polluting companies and market speculators. So-called "carbon cowboys" - unscrupulous entrepreneurs who attempt to acquire rights to carbon in rainforest for small-scale projects- have signed on indigenous communities to unfair contracts, often with a view to on-selling the rights to investors for a quick profit. In 2012 an Australian businessman operating in Peru was revealed to have signed 200-year contracts with an Amazon tribe, the Yagua, many members of which are illiterate, giving him a 50 per cent share in their carbon resources. The contracts allow him to establish and control timber projects and palm oil plantations in Yagua rainforest. This risk is largely negated by the focus on national and subnational REDD-plus programs, and by government ownership of these initiatives.

There are risks that the local inhabitants and the communities that live in the forests will be bypassed and that they won't be consulted and so they won't actually receive any revenues. Fair distribution of REDD-plus benefits will not be achieved without a prior reform in forest governance and more secure tenure systems in many countries.

The UNFCCC has repeatedly called for full and effective participation of Indigenous Peoples and local communities without becoming any more specific. The ability of local communities to effectively contribute to REDD-plus field activities and the measurement of forest properties for estimating reduced emissions and enhanced emissions of greenhouse gases has been clearly demonstrated in various countries.

In some project-based REDD-plus, disreputable companies have taken advantage of low governance.

Indigenous People

Safeguard: Respect for the knowledge and rights of indigenous peoples and members of local communities, by taking into account relevant international obligations, national circumstances and laws, and noting that the United Nations General Assembly has adopted the United Nations Declaration on the Rights of Indigenous Peoples; Safeguard: The full and effective participation of relevant stakeholders, in particular indigenous peoples and local communities, in the REDD-plus actions and when developing and implementing national strategies or action plans.

Indigenous peoples are important stakeholders in as they typically live inside forest areas or have their livelihoods (partially) based on exploitation of forest resources. The International Indigenous Peoples Forum on Climate Change (IIPFCC) was explicit at the Bali climate negotiations in 2007:

> "REDD/ REDD-plus will not benefit Indigenous Peoples, but in fact will result in more violations of Indigenous Peoples' rights. It will increase the violation of our human rights, our rights to our lands, territories and resources, steal

> our land, cause forced evictions, prevent access and threaten indigenous agricultural practices, destroy biodiversity and cultural diversity and cause social conflicts. Under REDD/REDD-plus, states and carbon traders will take more control over our forests."

Some claim putting a commercial value on forests neglects the spiritual value they hold for Indigenous Peoples and local communities.

Indigenous Peoples protested in 2008 against the United Nations Permanent Forum on Indigenous Issues final report on climate change and a paragraph that endorsed REDD-plus; this was captured in a video entitled "the 2nd May Revolt". However, these protests have largely disappeared in recent years. Indigenous people sit as permanent representatives on many multinational and national REDD-plus bodies, though there is always room for improvement.

Indigenous Peoples groups in Panama broke off their collaboration with the national UN-REDD Programme in 2012 over allegations of a failure of the government to properly respect the rights of the indigenous groups.

Some grassroots organizations are working to develop REDD-plus activities with communities and developing benefit-sharing mechanisms to ensure REDD-plus funds reach rural communities as well as governments. Examples of these include Plan Vivo projects in Mexico, Mozambique and Cameroon; and Carbonfund.org Foundation's VCS and CCBS projects in the state of Acre, Brazil.

REDD-plus in the Carbon Market

When REDD-plus was first discussed by the UNFCCC, no indication was given of the positive incentives that would support developing countries in their efforts to implement REDD-plus to reduce emissions and enhance removals of greenhouse gases from forests. In the absence of guidance from the COP, two options were debated by the international community at large:

- A market-based approach.
- A fund-based approach where Annex I countries would deposit substantial amounts of money into a fund administered by some multi-lateral entity.

Under the market-based approach, REDD-plus would act as an "offset scheme" in which verified results-based actions translate into some form of carbon credits, more-or-less analogous to the market for Certified Emission Reductions (CER) under the CDM of the Kyoto Protocol. Such carbon credits could then offset emissions in the country or company of the buyer of the carbon credits. This would require Annex I countries to agree to deeper cuts in emissions of greenhouse gases in order to create a market for the carbon credits from REDD-plus, which is unlikely to happen soon given the current state of negotiations in the COP, but even then there is the fear that the market will be flooded with carbon credits, depressing the price to levels where REDD-plus is no longer an economically viable option. Some developing countries, such as Brazil and China, maintain that developed countries must commit to real emissions reductions, independent of any offset mechanism.

Since COP-17, however, it has become clear that the REDD-plus may be financed by a variety of sources, market and non-market. The newly established Green Climate Fund already is supporting phase 1 and 2 REDD-plus programs, and is finalizing rules to allow disbursement of result-based finance to developing countries that submit verified reports of emission reductions and enhanced removals of greenhouse gases.

Top-down Design by Large International Institutions vs. Bottom-up Grassroots Coalitions

While the COP decisions emphasize national ownership and stakeholder consultation, there are concerns that some of the larger institutional organizations are driving the process, in particular outside of the *one Party, one vote* realm of multi-lateral negotiations under the UNFCCC. For example, the World Bank and the UN-REDD Programme, the two largest sources of funding and technical assistance for readiness activities and therefore unavoidable for most developing countries, place requirements upon recipient countries that are arguably not mandated or required by the COP decisions. A body of research suggests that, at least as of 2016, REDD-plus as a global architecture has only had a limited effect on local political realities, as pre-existing entrenched power dynamics and incentives that promote deforestation are not easily changed by the relatively small sums of money that REDD-plus has delivered to date. In addition, issues like land tenure that fundamentally determine who makes decisions about land use and deforestation have not been adequately addressed by REDD-plus, and there is no clear consensus on how complex political issues like land tenure can be easily resolved to favor standing forests over cleared forests through a relatively top-down mechanism like REDD-plus.

While a single, harmonized, global system that accounts for and rewards emissions reductions from forests and land use has been elusive, diverse context-specific projects have emerged that support a variety of activities including community-based forest management, enforcement of protected areas, sustainable charcoal production, and agroforestry. Although it is not clear whether these diverse projects are genuinely different from older integrated conservation and development initiatives that pre-date REDD-plus, there is evidence that REDD-plus has altered global policy conversations, possibly elevating issues like indigenous peoples' land rights to higher levels, or conversely threatening to bypass safeguards for indigenous rights. Debate surrounding these issues is ongoing.

Although the World Bank declares its commitment to fight against climate change, many civil society organisations and grassroots movements around the world view with scepticism the processes being developed under the various carbon funds. Among some of the most worrying reasons are the weak (or inexistent) consultation processes with local communities; the lack of criteria to determine when a country is ready to implement REDD-plus projects (readiness); the negative impacts such as deforestation and loss of biodiversity (due to fast agreements and lack of planning); the lack of safeguards to protect Indigenous Peoples' rights; and the lack of regional policies to stop deforestation. A growing coalition of civil society organization, social movement, and other actors critical of REDD-plus emerged between 2008 and 2011, criticizing the mechanism on climate justice grounds. During the UN climate negotiations in Copenhagen and Cancun strong civil society and social movements coalitions formed a strong front to fight the World Bank out of the climate. However, this concern has largely died down as the World Bank initiatives have been more full developed, and some of these same actors are now participating in implementation of REDD-plus.

ITTO has been criticized for appearing to support above all the inclusion of forest extraction inside REDD-plus under the guise of "sustainable management" in order to benefit from carbon markets while maintaining business-as-usual.

Forest Restoration

In the 1980s, conservation organizations warned that, once destroyed, tropical forests could never be restored. Thirty years of restoration research now challenge this:

- This site in Doi Suthep-Pui National Park, N. Thailand was deforested, over-cultivated and then burnt. The black tree stump was one of the original forest trees. Local people teamed up with scientists to repair their watershed.

- Fire prevention, nurturing natural regeneration and planting framework tree species resulted in trees growing above the weed canopy within a year.

- After 12 years, the restored forest overwhelmed the black tree stump.

Forest restoration is defined as "actions to re-instate ecological processes, which accelerate recovery of forest structure, ecological functioning and biodiversity levels towards those typical of climax forest" i.e. the end-stage of natural forest succession. Climax forests are relatively stable ecosystems that have developed the maximum biomass, structural complexity and species diversity that are possible within the limits imposed by climate and soil and without continued disturbance from humans. Climax forest is therefore the target ecosystem, which defines the ultimate aim of forest restoration. Since climate is a major factor that determines climax forest composition, global climate change may result in changing restoration aims.

Forest restoration is a specialized form of reforestation, but it differs from conventional tree plantations in that its primary goals are biodiversity recovery and environmental protection.

Forest restoration may include simply protecting remnant vegetation (fire prevention, cattle exclusion etc.) or more active interventions to accelerate natural regeneration, as well as tree planting and sowing seeds (direct seeding) of species characteristic of the target ecosystem. Tree species planted (or encouraged to establish) are those that are typical of, or provide a critical ecological function in, the target ecosystem. However, wherever people live in or near restoration sites, restoration projects often include economic species amongst the planted trees, to yield subsistence or cash-generating products.

Forest restoration is an inclusive process, which depends on collaboration among a wide range of stakeholders including local communities, government officials, non-government organizations, scientists and funding agencies. Its ecological success is measured in terms of increased biological diversity, biomass, primary productivity, soil organic matter and water-holding capacity, as well as the return of rare and keystone species, characteristic of the target ecosystem. Economic indices of success include the value of forest products and ecological services generated (e.g. watershed protection, carbon storage etc.), which ultimately contribute towards poverty reduction. Payments for such ecological services (PES) and forest products can provide strong incentives for local people to implement restoration projects.

Opportunities for Forest Restoration

Demonstration forest restoration plot.

Forest restoration is appropriate wherever biodiversity recovery is one of the main goals of reforestation, such as for wildlife conservation, environmental protection, eco-tourism or to supply a wide variety of forest products to local communities. Forests can be restored in a wide range of circumstances, but degraded sites within protected areas are a high priority, especially where some climax forest remains as a seed source within the landscape. Even in protected areas, there are often large deforested sites: logged over areas or sites formerly cleared for agriculture. If protected areas are to act as Earth's last wildlife refuges, restoration of such areas will be needed.

Many restoration projects are now being implemented under the umbrella of "forest landscape restoration" (FLR), defined as a "planned process to regain ecological integrity and enhance human well-being in deforested or degraded landscapes". FLR recognizes that forest restoration has social and economic functions. It aims to achieve the best possible compromise between meeting both conservation goals and the needs of rural communities. As human pressure on landscapes increases, forest restoration will most commonly be practiced within a mosaic of other forms of forest management, to meet the economic needs of local people.

A recent focal area for forest restoration efforts is within the urban context, where both people and biodiversity will benefit, however this context presents unique challenges.

Natural Regeneration

Tree planting is not always essential to restore forest ecosystems. A lot can be achieved by studying how forests regenerate naturally, identifying the factors that limit regeneration and devising methods to overcome them. These can include weeding and adding fertilizer around natural tree seedlings, preventing fire, removing cattle and so on. This is "accelerated" or "assisted" natural regeneration. It is simple and cost-effective, but it can only operate on trees that are already present, mostly light-loving pioneer species. Such tree species are not usually those that comprise climax forests, but they can foster recolonization of the site by shade-tolerant climax forest tree species, via natural seed dispersal from remnant forest. Because this is a slow process, biodiversity recovery can usually be accelerated by planting some climax forest tree species, especially large-seeded, poorly dispersed species. It is not feasible to plant all the tree species that may have formerly grown in the original primary forest and it is usually unnecessary to do so, if the framework species method can be used.

Post-fire Regeneration

In large parts of the world, forest fires cover a heavy toll on forests. That can be because of provoked deforestation in order to substitute forests by crop areas, or in dry areas, because of wild fires occurring naturally or intentionally. A whole topic of forest landscape restoration in linked to this particular problem, as in many cases, the net loss of ecosystem value is very high and can open the drop to an accelerated further degradation of the soil conditions through erosion and desertification. This indeed has dire consequences on both the quality of the habitats and their related fauna. Nevertheless, in some specific cases, wild fires do actually allow to increase the biodiversity index of the burnt area, in which case the Forest Restoration Strategies tend to look for a different land-use.

Forest Landscape Restoration

Forest landscape restoration (FLR) is defined as "a planned process to regain ecological integrity and enhance human well-being in deforested or degraded landscapes". It comprises tools and procedures to integrate site-level forest restoration actions with desirable landscape-level objectives, which are decided upon via various participatory mechanisms among stakeholders. The concept has grown out of collaboration among some of the world's major international conservation organizations including the International Union for Conservation of Nature (IUCN), the World Wide Fund for Nature (WWF), the World Resources Institute and the International Tropical Timber Organization (ITTO).

Aims

The concept of FLR was conceived to bring about compromises between meeting the needs of both humans and wildlife, by restoring a range of forest functions at the landscape level. It includes actions to strengthen the resilience and ecological integrity of landscapes and thereby keep future management options open. The participation of local communities is central to the concept, because they play a critical role in shaping the landscape and gain significant benefits from restored forest resources. Therefore, FLR activities are inclusive and participatory.

Desirable Outcomes

The desirable outcomes of an FLR program usually comprise a combination of the following, depending on local needs and aspirations:

- Identification of the root causes of forest degradation and prevention of further deforestation.
- Positive engagement of people in the planning of forest restoration, resolution of land-use conflicts and agreement on benefit-sharing systems.
- Compromises over land-use trade-offs that are acceptable to the majority of stakeholders.
- A repository of biological diversity of both local and global value.
- Delivery of a range of utilitarian benefits to local communities including:
    - A reliable supply of clean water.
    - Environmental protection particularly watershed services (e.g. reduced soil erosion, lower landslide risk, flood/drought mitigation etc.).
    - A sustainable supply of a diverse range of forest products including foods, medicines, firewood etc.
    - Monetary income from various sources e.g. ecotourism, carbon trading via the REDD-plus mechanism and from payments for other environmental services (PES).

Activities

FLR combines several existing principles and techniques of development, conservation and natural resource management, such as landscape character assessment, participatory rural appraisal, adaptive management etc. within a clear and consistent evaluation and learning framework. An FLR program may comprise various forestry practices on different sites within the landscape, depending on local environmental and socioeconomic factors. These may include protection and management of secondary and degraded primary forests, standard forest restoration techniques such as "assisted" or "accelerated" natural regeneration (ANR) and the planting of framework tree species to restore degraded areas, as well as conventional tree plantations and agroforestry systems to meet more immediate monetary needs.

The IUCN hosts the Global Partnership on Forest Landscape Restoration, which co-ordinates development of the concept around the world.

In 2014, the Food and Agricultural Organization of the United Nations established the Forest and Landscape Restoration Mechanism. The Mechanism supports countries to implement FLR as a contribution to achieving the Bonn Challenge - the restoration of 150 million hectare of deforested and degraded lands by 2020 - and the Convention on Biological Diversity Aichi Biodiversity Targets - related to ecosystem conservation and restoration.

Afforestation

Afforestation is the process of introducing trees and tree seedlings to an area that has previously not been forested. Afforestation can be done through tree planting and seeding naturally or artificially. Similarly, reforestation can be considered a form of afforestation because like afforestation, reforestation is the alteration of a non-forested area to a forested area through tree planting and seeding. The difference is that reforestation is the restoration of an area that has been deforested.

Purposes of Afforestation

There are many reasons for restoring an area via afforestation and reforestation; however they vary based on the area. Generally it is done for either conservation or industrial-commercial purposes:

- Conservational afforestation and reforestation is done in the best interest of the ecosystem. Its purpose is to restore an area that has been destroyed due to previous overuse of the land or to reduce the amount of erosion in the soil in an area and establish a more fertile soil base.
- Industrial-commercial afforestation and reforestation is done to maintain a good output of wood for pulp and timber demands in a specific area.

Impacts of Afforestation

Afforestation, reforestation and other forms of conservational forestry methods are often thought to be used for stopping the effects of climate change by reducing atmospheric carbon. The issue is which type of tree sequesters the most amount of carbon and does this have a positive or negative effect if any at all? The tree types that are widely discussed are old growth vs. young-growth forests. Old growth forests are said to have massive carbon storage capabilities however these trees are capturing carbon incredibly slowly or unable to capture any more. With reforestation, this process is 'solved' by cutting down old growth and planting young-growth trees. Unfortunately, when an old growth forest is cut down, a lot of the stored carbon gets released into the atmosphere thus preventing a net positive effect. Afforestation is being thought of as a solution to the reforestation issue. By planting a new young-growth forest in an area that hasn't previously had a forested, this could be a viable option for sequestering more carbon from the atmosphere. However, planting trees in an area previously unforested could impact the original ecosystem negatively such as reducing soil moisture in an area and forcing many species to leave their preferred habitats.

Issues affecting Afforestation

Some issues that can arise from afforestation and reforestation are climate change, anthropogenic changes to landscape and increased wood demand. Climate change can have some major effects on the growth and health on newly planted trees. Drought and major temperature fluctuations can cause serious seedling mortality to an area that has been newly planted, ultimately slowing down the restoration process. To reduce the effects of climate change on the restoration process, drought and shade tolerant species are used.

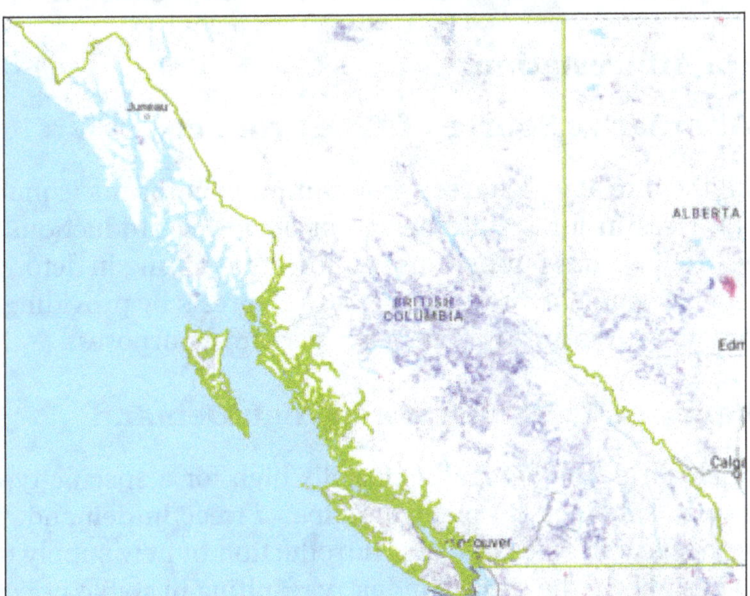

In 2001 the total tree cover loss was 200 000 ha.

Increasing wood demand is another problem. The rate at which trees are being planted is much slower than the rate of those being cut down for production. The trees need time to grow and without a properly timed planting seasons going hand in hand with deforestation, there can be a potential stasis for the ever increasing wood demand.

Climate change is also changing the success of pests like the mountain pine beetle. For example, in North America, as temperatures rise in the west and winters are becoming warmer, the pine beetle populations aren't dying out as they are supposed to in the winter. The difference between tree cover gain and loss in British Columbia from 2001 to 2014 taken from the global forest watch interactive map. The pink dots represent tree cover loss and the blue dots represent tree forest gain.

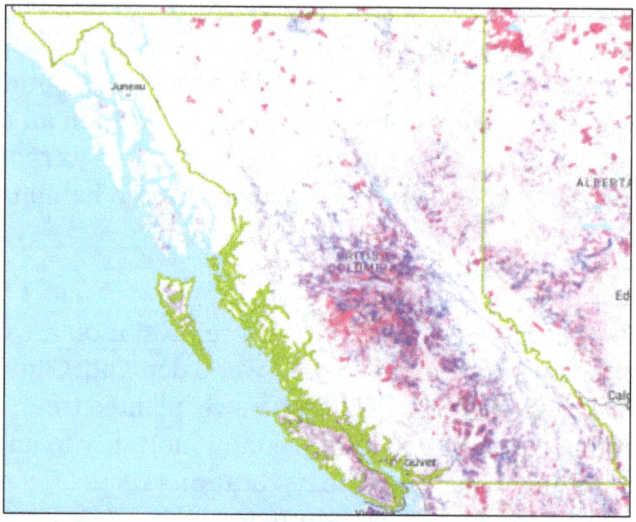

From 2001 to 2014 the total tree cover loss was 5 000 000 ha.

Importance of Afforestation

Provide an Alternative Source of Tree Products

In today's world, the rate at which trees grow naturally in forests is much slower than the rate at which trees in forests are being cut down for production. The increased demand for tree products has put pressure on forests resulting in deforestation. Afforestation is helping alleviate the pressure on natural forests by providing an alternative source of tree products. It is mostly done for commercial purposes.

Increased Supply of Trees that are in High Demand

In the commercial world, the demand is usually high for a specific type of trees. Afforestation allows stakeholders to plant the type of trees in demand, promoting the fast propagation of specific types of trees. Introduction of new supply of trees, for instance, can help to stabilize the cottage industry resulting in stable prices and business reliability.

Protection of Natural Forests

Protecting natural forests without providing alternative sources of tree products is very difficult and unsustainable. Besides, logging is more common in places where there are no alternative sources of tree products.

However, when commercial tree users have alternative sources of supply, they can support the administrative and policy efforts in protecting forests, resulting in more sustainable initiatives. Protecting natural forests also leads to other benefits that come with increased tree forest cover like preservation of catchments, wetlands, and river-side zones.

Environmental Benefits

Planting trees in a barren land is in some occasions more beneficial than planting trees in a depleted or depleting forest. Planting trees in a depleted or recovering forests helps to restore the ecosystem of the area, but planting trees in a barren land helps to create a new ecosystem. Most governments and institutions have used afforestation to help revert the arid and semi-arid areas to productive areas. It also helps improve the appearance of barren lands to picturesque areas.

Value Addition

An easy way to transform a barren land is planting trees. Land with trees is more attractive and valuable than barren land. Whether one is seeking to develop or sell property, planting trees is a sure way of increasing the property's value.

Various Advantages of Afforestation

Constant Supply of Forest Products

Planting trees in the land that was not previously productive ensures that the supply of essential products does not stop. Relying on natural forests can be highly unreliable. It is also not predictive due to government and environmental policies.

Forest harvesting is highly regulated. It is usually common for governments to put a ban on forest harvesting, which can be detrimental for a business that fully relies on forest supplies. Afforestation provides people and businesses with reliable alternative sources of forest supplies. For example, a company can easily and reliably project its annual sales and production when it has its own land of trees, unlike when it sources its products from natural forests.

Creating new forests also provide the community with fodder, fruits, firewood, and other valuable resources necessary for humans. In addition to making the resources available, afforestation also ensures that the resources are sustainable for the future.

Prevents Soil Erosion

Soil erosion is a common problem in barren lands. Barren land is not the best place to be on a windy day. Barren lands experience strong winds that carry large particles of soil causing soil erosion while also negatively affecting air quality. A great example is the dust bowl that was highly associated with intensive farming and large tracts of barren lands.

Barren lands are also subject to flooding when it rains, resulting in the topsoil being transported to rivers and water sources in the area. Afforestation helps to address all the issues of soil erosion and barren land at the same time. Trees act as wind barriers thereby weakening the speed of the wind and reducing its impact and ability to carry big soil particles. Tree roots also hold the soil together tightly making sure that the soil stays in place during flooding.

Tree leaves and branches also help to reduce the impact of raindrops on the ground preventing water drops erosion. Keeping the soil in place ensures that farmers have fertile soils to grow crops. Soil that held tightly is also less prone to landslides, especially in hilly and mountain regions.

Stabilizes the Climate

Planting trees is a proven way of transforming arid and semiarid areas. Places with more trees experience frequent rains, and thus, planting trees is a perfect way to stabilize climate. Trees also reduce the greenhouse effect preventing warmer temperatures as they are the world's only natural air conditioners.

Potential of Reversing Global Warming and Climate Change

Global warming is a major concern for global leaders and environmentalists. As such, more and more people are being encouraged to plant trees in their gardens, and afforestation is one of the recommended ways of addressing global warming.

Trees help reverse the greenhouse effect through the process of photosynthesis as they act as carbon sinks. In other words, establishing new forests create new carbon dioxide holding areas – reducing the carbon dioxide in the environment. The overall outcome is the reduction of the effect of global warming.

Better Quality Air

Trees play an important role in purifying the air. People in areas with trees are less likely to suffer from air-related conditions. Trees purify carbon dioxide and offer oxygen through photosynthesis. In the last decades, people have been producing more and more carbon dioxide through their activities like driving, burning fossils, and industrial activities, resulting in more carbon dioxide in the air.

In addition to producing carbon dioxide, human activities like construction have led to the diminishing of forests. The imbalance has led to more toxins in the air. In addition to carbon dioxide and greenhouse emissions, human activities like construction and driving create dust particles in the air affecting the air quality. Trees not only purify carbon dioxide, but they also purify greenhouse emissions. They also trap soil particles in the air resulting in better air quality.

Improve Watershed

Without clean water, the earth cannot support lives. Humans and animals rely on clean water for their survival and hence the need to protect the watershed. Although most efforts are on restoring the existing watersheds, afforestation plays a major role in creating new watersheds. People living in arid and semiarid regions are the most affected when it comes to lack of clean water for consumption and household use.

Places with forests have several sources of water as opposed to places without trees where they mostly rely on floodwater conservation. Afforestation can help improve the watershed of arid and semiarid areas. Trees help to bring more rain to the region. Trees also ensure that the water is better held in the soil, improving the underground water table.

Preservation of Wildlife

When forests diminish, the wildlife also diminishes. Wild animals suffer most when humans invade forests and their other habitats. Today, most of the places where humans have built cities and estates used to be grazing and dwelling places for wildlife. As human activities continue to expand, the number of wildlife living in unprotected areas has dwindled.

Furthermore, unless drastic measures are embraced, most of these animals will become extinct. Afforestation helps to ensure that there are enough forests for wildlife to thrive in. Those animals pushed from their natural habitats by human activities can relocate to the new forests. For this reason, afforestation can aid in protecting wild animals.

Provision of Employment and Economic Opportunities

The timber industry is a big employer. In the construction industry, timber products are basic components of every building. However, before they become ready for final use in the construction site, they undergo detailed processing. For example, before hardboards can be used in the construction of drawers, they have to be processed from round timber.

The process involves different stages that require a unique set of skills prompting the need to hire different workers. Furthermore, since afforestation is the process of turning barren lands into forests, they require a huge number of labourers to perform different tasks that include soil digging, sowing seeds, watering plants, and harvesting of trees. Most of the people employed in these lands are previously unemployed and usually poor.

Reforestation

Tropical tree nursery.

A 15-year-old reforested plot of land.

Reforestation is the natural or intentional restocking of existing forests and woodlands (forestation) that have been depleted, usually through deforestation. Reforestation can be used to rectify or improve the quality of human life by soaking up pollution and dust from the air, rebuild natural habitats and ecosystems, mitigate global warming since forests facilitate bio sequestration of atmospheric carbon dioxide and harvest for resources, particularly timber, but also non-timber forest products.

A 21-year-old plantation of red pine in Southern Ontario.

Management

A debated issue in managed reforestation is whether or not the succeeding forest will have the same biodiversity as the original forest. If the forest is replaced with only one species of tree and all other vegetation is prevented from growing back, a monoculture forest similar to agricultural crops would be the result. However, most reforestations involve the planting of different selections of seedlings taken from the area, often of multiple species. Another important factor is the natural regeneration of a wide variety of plant and animal species that can occur on a clear cut. In some areas the suppression of forest fires for hundreds of years has resulted in large single aged and single species forest stands. The logging of small clear cuts and/or prescribed burning actually increases the biodiversity in these areas by creating a greater variety of tree stand ages and species.

For Harvesting

Reforestation need not be only used for recovery of accidentally destroyed forests. In some countries, such as Finland, many of the forests are managed by the wood products and pulp and paper industry. In such an arrangement, like other crops, trees are planted to replace those that have been cut. In such circumstances, the industry can cut the trees in a way to allow easier reforestation. The wood products industry systematically replaces many of the trees it cuts, employing large numbers of summer workers for tree planting work. For example, in 2010, Weyerhaeuser reported planting 50 million seedlings. However replanting an old-growth forest with a plantation is not replacing the old with the same characteristics in the new.

In just 20 years, a teak plantation in Costa Rica can produce up to about 400 m^3 of wood per hectare. As the natural teak forests of Asia become more scarce or difficult to obtain, the prices commanded by plantation-grown teak grows higher every year. Other species such as mahogany grow more slowly than teak in Tropical America but are also extremely valuable. Faster growers include pine, eucalyptus, and Gmelina.

Reforestation, if several indigenous species are used, can provide other benefits in addition to financial returns, including restoration of the soil, rejuvenation of local flora and fauna, and the capturing and sequestering of 38 tons of carbon dioxide per hectare per year.

The reestablishment of forests is not just simple tree planting. Forests are made up of a community of species and they build dead organic matter into soils over time. A major tree-planting program could enhance the local climate and reduce the demands of burning large amounts of fossil fuels for cooling in the summer.

For Climate Change Mitigation

Forests are an important part of the global carbon cycle because trees and plants absorb carbon dioxide through photosynthesis. By removing this greenhouse gas from the air, forests function as terrestrial carbon sinks, meaning they store large amounts of carbon. At any time, forests account for as much as double the amount of carbon in the atmosphere. Even as more anthropogenic carbon is produced, forests remove around three billion tons of anthropogenic carbon every year. This amounts to about 30% of all carbon dioxide emissions from fossil fuels. Therefore, an increase in the overall forest cover around the world would tend to mitigate global warming.

There are four major strategies available to mitigate carbon emissions through forestry activities: increase the amount of forested land through a reforestation process; increase the carbon density of existing forests at a stand and landscape scale; expand the use of forest products that will sustainably replace fossil-fuel emissions; and reduce carbon emissions that are caused from deforestation and degradation.

Achieving the first strategy would require enormous and wide-ranging efforts. However, there are many organizations around the world that encourage tree-planting as a way to offset carbon emissions for the express purpose of fighting climate change. For example, in China, the Jane Goodall Institute, through their Shanghai Roots and Shoots division, launched the Million Tree Project in Kulun Qi, Inner Mongolia to plant one million trees to stop desertification and help curb climate change. China has used 24 billion metres squared of new forest plantation and natural forest regrowth to offset 21% of Chinese fossil fuel emissions in 2000. In Java, Indonesia each newlywed couple is to give whoever is sermonizing their wedding 5 seedlings to combat global warming. Each couple that wishes to have a divorce has to give 25 seedlings to whoever divorces them. By reforestation and environmental conservation, Costa Rica doubled its forest cover in 30 Years. It has a long-standing commitment to the environment. The country is now one of the leaders of sustainability, biodiversity, and other protections. Costa Rica has pioneered the development of payments for environmental services. The system of environmental protection provid grants for environmental services. This system is not just advanced for its time but is also unparalleled in the world. It received great international attention. Also the country has established programs to compensate landowners for reforestation. These payments are funded through international donations and nationwide taxes.

The second strategy has to do with selecting species for tree-planting. In theory, planting any kind of tree to produce more forest cover would absorb more carbon dioxide from the atmosphere. On the other hand, a genetically modified tree specimen might grow much faster than any other regular tree. Some of these trees are already being developed in the lumber and biofuel industries. These fast-growing trees would not only be planted for those industries but they can also be planted to help absorb carbon dioxide faster than slow-growing trees.

Extensive forest resources placed anywhere in the world will not always have the same impact. For example, large reforestation programs in boreal or subarctic regions have a limited impact on climate mitigation. This is because it substitutes a bright snow-dominated region that reflects the sunlight with dark forest canopies. On the other hand, a positive example would be reforestation projects in tropical regions, which would lead to a positive biophysical change such as the formation of clouds. These clouds would then reflect the sunlight, creating a positive impact on climate mitigation.

There is an advantage to planting trees in tropical climates with wet seasons. In such a setting, trees have a quicker growth rate because they can grow year-round. Trees in tropical climates have, on average, larger, brighter, and more abundant leaves than non-tropical climates. A study of the girth of 70,000 trees across Africa has shown that tropical forests are soaking up more carbon dioxide pollution than previously realized. The research suggests almost one fifth of fossil fuel emissions are absorbed by forests across Africa, Amazonia and Asia. Simon Lewis, a climate expert at the University of Leeds, who led the study, said: "Tropical forest trees are absorbing about 18% of the carbon dioxide added to the atmosphere each year from burning fossil fuels, substantially buffering the rate of change."

It is also important to deal with the rate of deforestation. At this point, there are 13 billion metres squared of tropical regions that are deforested every year. There is potential for these regions to reduce rates of deforestation by 50% by 2050, which would be a huge contribution to stabilize the global climate.

Incentives

Some incentives for reforestation can be as simple as a financial compensation. Streck and Scholz explain how a group of scientists from various institutions have developed a compensated reduction of deforestation approach which would reward developing countries that disrupt any further act of deforestation. Countries that participate and take the option to reduce their emissions from deforestation during a committed period of time would receive financial compensation for the carbon dioxide emissions that they avoided. To raise the payments, the host country would issue government bonds or negotiate some kind of loan with a financial institution that would want to take part in the compensation promised to the other country. The funds received by the country could be invested to help find alternatives to the extensive cut down of forests. This whole process of cutting emissions would be voluntary, but once the country has agreed to lower their emissions they would be obligated to reduce their emissions. However, if a country was not able to meet their obligation, their target would get added to their next commitment period.

Related Concepts

A similar concept, afforestation, refers to the process of restoring and recreating areas of woodlands or forests that may have existed long ago but were deforested or otherwise removed at some point in the past or lacked it naturally (e.g., natural grasslands). Sometimes the term "re-afforestation" is used to distinguish between the original forest cover and the later re-growth of forest to an area. Special tools, e.g. tree planting bars, are used to make planting of trees easier and faster.

Another alternative strategy, proforestation, is similar as it can be used to counteract the negative environmental and ecological effects of deforestation through growing an existing forest intact to its full ecological potential.

Criticisms

Reforestation competes with other land uses, such as food production, livestock grazing and living space, for further economic growth. Reforestation often has the tendency to create large fuel loads, resulting in significantly hotter combustion than fires involving low brush or grasses. Reforestation can divert large amounts of water from other activities. Reforesting sometimes results in extensive canopy creation that prevents growth of diverse vegetation in the shadowed areas and generating soil conditions that hamper other types of vegetation. Trees used in some reforesting efforts (e.g., Eucalyptus globulus) tend to extract large amounts of moisture from the soil, preventing the growth of other plants.

There is also the risk that, through a forest fire or insect outbreak, much of the stored carbon in a reforested area could make its way back to the atmosphere. Reduced harvesting rates and fire suppression have caused an increase in the forest biomass in the western United States over the past century. This causes an increase of about a factor of four in the frequency of fires due to longer and hotter dry seasons.

Protective Measures for Forest Conservation

For the conservation of forests, following steps can be taken:

- Conservation of forest is a national problem so it must be tackled with perfect coordination between forest department and other departments.
- People's participation in the conservation of forests is of vital importance. So, we must get them involved in this national task.
- The cutting of trees in the forests must be stopped at all costs.
- Afforestation or special programmes should be launched on grand scale.
- Celebrations of all functions, festivals should precede with tree-plantation.
- Cutting of timber and other forest produce should be restricted.
- Grasslands should be regenerated.
- Several centres of excellence have been setup and awards should be instituted.

References

- Forest-conservation-useful-methods-for-forest-conservation, environment-forest-25277: you-rarticlelibrary.com, Retrieved 13 January, 2020

- Streck, C.; S.M. Scholz (2006). "The role of forests in global climate change: whence we come and where we go". International Affairs. 82 (5): 861–879. doi:10.1111/j.1468-2346.2006.00575.x
- Afforestation, encyclopedia: energyeducation.ca, Retrieved 03 June, 2020
- Chazdon, Robin L.; Wilson, Sarah; Strassburg, Bernardo B. N.; Reid, J. Leighton; Goetz, Scott; Aronson, James; Baccini, Alessandro; Zambrano, Angelica M. Almeyda; Barros, Felipe S. M. (2019-07-01). "Global restoration opportunities in tropical rainforest landscapes". Science Advances. 5 (7): eaav3223. doi:10.1126/sciadv.aav3223. ISSN 2375-2548
- Importance-advantages-afforestation: conserve-energy-future.com, Retrieved 28 August, 2020
- 9-simple-methods-of-forest-conservation, environment-4006: preservearticles.com, Retrieved 05 March, 2020
- Philip Fearnside (2000). "Global warming and tropical land-use change: Greenhouse gas emissions from biomass burning, decomposition and soils in forest conversion, shifting cultivation and secondary vegetation". Climatic Change. 46: 115–158. doi:10.1023/a:1005569915357

Permissions

All chapters in this book are published with permission under the Creative Commons Attribution Share Alike License or equivalent. Every chapter published in this book has been scrutinized by our experts. Their significance has been extensively debated. The topics covered herein carry significant information for a comprehensive understanding. They may even be implemented as practical applications or may be referred to as a beginning point for further studies.

We would like to thank the editorial team for lending their expertise to make the book truly unique. They have played a crucial role in the development of this book. Without their invaluable contributions this book wouldn't have been possible. They have made vital efforts to compile up to date information on the varied aspects of this subject to make this book a valuable addition to the collection of many professionals and students.

This book was conceptualized with the vision of imparting up-to-date and integrated information in this field. To ensure the same, a matchless editorial board was set up. Every individual on the board went through rigorous rounds of assessment to prove their worth. After which they invested a large part of their time researching and compiling the most relevant data for our readers.

The editorial board has been involved in producing this book since its inception. They have spent rigorous hours researching and exploring the diverse topics which have resulted in the successful publishing of this book. They have passed on their knowledge of decades through this book. To expedite this challenging task, the publisher supported the team at every step. A small team of assistant editors was also appointed to further simplify the editing procedure and attain best results for the readers.

Apart from the editorial board, the designing team has also invested a significant amount of their time in understanding the subject and creating the most relevant covers. They scrutinized every image to scout for the most suitable representation of the subject and create an appropriate cover for the book.

The publishing team has been an ardent support to the editorial, designing and production team. Their endless efforts to recruit the best for this project, has resulted in the accomplishment of this book. They are a veteran in the field of academics and their pool of knowledge is as vast as their experience in printing. Their expertise and guidance has proved useful at every step. Their uncompromising quality standards have made this book an exceptional effort. Their encouragement from time to time has been an inspiration for everyone.

The publisher and the editorial board hope that this book will prove to be a valuable piece of knowledge for students, practitioners and scholars across the globe.

Index

A
Agroforestry, 5, 55, 86-89, 91, 93-96, 98, 102, 194, 218, 224
Analog Forestry, 78-79, 85
Artificial Regeneration, 51, 84, 107-108, 117, 137
Assisted Natural Regeneration, 105, 198

B
Balsam Fir, 34, 38, 107, 113
Bark Beetles, 62, 122
Beta Diversity, 14, 27
Biodiversity Patterns, 14, 27
Biofuels, 23
Biome, 11-13, 15-16, 24, 26, 28-29, 31-33, 35-36, 39, 42, 44, 214
Biotic Species, 17
Buffer Strips, 7

C
Carbon Sequestration, 79, 82, 166-167, 199-200
Cloud Forests, 25, 192, 194
Community Forestry, 5, 79-83
Coniferous Forests, 3, 10, 28, 35, 55-56, 59-60
Continuous Cover Forestry, 83-85, 120

D
Deciduous Forest, 1, 15-16, 26, 187
Deforestation, 17, 21-23, 27, 66, 156-157, 160, 168, 175, 194, 197, 200-206, 210, 214, 218-219, 222-223, 226, 230, 232-233

E
Ecoforestry, 74-76, 158
Evergreen Forest, 10

F
Flat Leaves, 15
Food Web, 15-16, 36
Forest Cover, 4, 12-13, 33, 58, 62, 162, 178, 183, 187, 189, 193, 196-198, 201, 227, 232-233
Forest Crop, 53, 111
Forest Degradation, 200-206, 210, 214-215, 223
Forest Farming, 86, 89-92
Forest Floor, 3, 12, 16, 18-20, 24, 32, 35, 38, 58-59, 106, 116, 123-124, 127-128, 149, 163
Forest Gardening, 91, 93-94
Forest Management, 7, 39, 45-50, 55, 63-70, 72, 74, 79-84, 90, 92, 96, 111, 113, 119, 131, 149, 158, 168, 173, 177-181, 189, 201, 210, 213, 218, 222
Forest Restoration, 96, 189, 196, 199, 219-224
Freshwater Swamp Forest, 10, 42-43

G
Geographical Variation, 26

H
Habitat Loss, 17, 159
Herbaceous Floras, 14
Hydrologic Cycle, 58

I
Intertropical Convergence Zone, 17-18

L
Legumes, 55
Lichen Woodland, 33, 42

M
Machine Planting, 53
Mangrove Forests, 11
Mineral Deposits, 21
Mycoforestry, 86, 96-98

N
National Forest Funds, 68, 85
Natural Ecosystem, 81
Natural Landscapes, 15
Natural Regeneration, 51, 76, 105-108, 112-113, 115, 120, 122, 124, 130, 136, 164, 198-199, 206, 220, 222, 224, 231

O
Organic Matter, 3, 54, 70, 87, 128, 193, 221, 231

P

Populus Tremula, 31

Precipitation, 1, 16, 31, 35, 37, 40-41, 58, 106, 186, 191-192, 194, 197

R

Regeneration Capacity, 65

Resource Inventory, 48, 92

Rhizobium, 52

Riparian Buffer, 98-101, 192, 194-195

Riparian Zones, 76, 101

S

Saprotrophic Fungi, 96-97

Savanna, 2

Semiarid Regions, 59, 229

Shrub Layer, 3, 16, 20, 95

Silvicultural System, 51, 115, 124, 131-133, 148, 158, 162

Soil Erosion, 21, 51, 60, 87-89, 159, 166, 180, 189-190, 193, 197, 223, 228

Soil Fertility, 87-89, 139, 168, 190

Soil Quality, 21

Solar Energy, 30, 87, 150

Species Composition, 1, 13, 70, 72, 85, 116, 126, 185, 191

Spruce Budworm, 41-42

Subarctic Climate, 29, 42

Subtropical Moist Broadleaf Forests, 23

Sustainable Forest, 3, 45, 64-68, 70, 80, 82, 92, 177-181, 183-184, 189, 191, 213

T

Taiga, 1-2, 28-40, 42, 157

Timber Cuttings, 48

Topography, 2, 32, 48, 131, 191

Transpiration, 2, 11, 35, 58, 150

Tree Canopy, 3, 12, 89

Tree Crown, 91, 148

Tropical Forests, 3, 18, 22, 44, 74, 177, 219, 233

Tropical Moist Forests, 18, 23

U

Understory Layer, 19

Urban Forestry, 64

W

Water Table, 43, 58, 88, 192, 194, 229

Watershed Management, 57, 190

Weathering Rock, 88

Wildlife Habitat, 6, 47-49, 90, 99, 103, 118, 126, 157

Wildlife Management, 56-57, 91

www.ingramcontent.com/pod-product-compliance
Lightning Source LLC
LaVergne TN
LVHW080627210926
848485LV00003B/220
* 9 7 8 1 6 3 9 8 9 2 1 1 2 *